AGRICULTURAL BIOTECHNOLOGY AND TRANSATLANTIC TRADE

REGULATORY BARRIERS TO GM CROPS

Agricultural Biotechnology and Transatlantic Trade

Regulatory Barriers to GM Crops

Grant E. Isaac
Associate Professor
Department of Management and Marketing
College of Commerce
University of Saskatchewan
Canada

CABI *Publishing*

CABI *Publishing* is a division of CAB *International*

CABI Publishing
CAB International
Wallingford
Oxon OX10 8DE
UK

Tel: +44 (0)1491 832111
Fax: +44 (0)1491 833508
Email: cabi@cabi.org
Web site: http://www.cabi-publishing.org

CABI Publishing
10 E 40th Street
Suite 3203
New York, NY 10016
USA

Tel: +1 212 481 7018
Fax: +1 212 686 7993
Email: cabi-nao@cabi.org

A catalogue record for this book is available from the British Library, London, UK

Library of Congress Cataloging-in-Publication Data
Isaac, Grant.
Agricultural biotechnology and transatlantic trade : regulatory barriers to GM crops / Grant E. Isaac.
p. cm.
Includes bibliographical references (p.).
ISBN 0-85199-580-2 (alk. paper)
1. Genetic engineering industry. 2. Genetically modified foods. 3. Food industry and trade. 4. Agricultural biotechnology. 5. International trade. I. Title.

HD9999.G452 I83 2002

2001043286

ISBN 0 85199 580 2

Typeset in 10/12pt Melior by Columns Design Ltd, Reading.
Printed and bound in the UK by Cromwell Press, Trowbridge.

Contents

List of Tables

List of Figures

Preface

This study was motivated by a simple research question: why are genetically modified (GM) agricultural crops approved as safe in North America (Canada and the USA) facing significant regulatory hurdles in gaining access to the European Union (EU)? At first glance, this seemed to be a classic trade problem with a classic remedy – the establishment of rules within the international trading regime to discipline those EU regulations preventing market access for North American GM crops. Yet very quickly it became clear that this was not a classic trade problem and as such the classic trade remedy was not a viable solution.

Instead, the development and commercialization of GM crops illustrates a complex challenge facing trade diplomacy – the challenge of regulatory regionalism created by social regulatory barriers. Social regulations associated with GM crops have been enacted in regulatory jurisdictions to ensure food safety and environmental protection and to meet moral, ethical and religious preferences. When the regulatory approaches within jurisdictions differ significantly, regulatory regionalism is created. For instance, while Canada and the USA have very similar approaches to regulating GM crops, these approaches differ significantly from the regulatory approach in the EU, resulting in transatlantic regulatory regionalism; GM crops approved as safe in North America have been delayed or denied market access in the EU. Therefore, the trade problem is not a classic one of disciplining border measures, such as tariffs and quantitative restrictions, but a complex and controversial one of reaching deep into a jurisdiction's regulatory competence and disciplining the jurisdiction's regulatory policies on crucial issues, such as food safety and environmental protection.

As border measures have come under greater trade discipline and domestic social regulations have emerged on the trade agenda, trade diplomacy has arrived at a crucial crossroads. The traditional integration approach of trade diplomacy, which continues to work well for dealing with border measures, does not effectively deal with social regulatory barriers to trade. It fails to acknowledge that social regulations are a function of endogenous political-economy factors that can differ substantially between jurisdictions, resulting, of course, in substantially different regulatory approaches. In this sense, regulatory development within a jurisdiction sets the parameters for the type of regulatory integration strategy pursued between jurisdictions. Imposing top-down regulatory integration rules upon those jurisdictions whose regulatory approaches are not considered compliant with the traditional trade approach will not facilitate regulatory integration. Instead, it will erode public support for trade diplomacy and marginalize it as a viable force in international integration. Indeed, protests at the World Trade Organization's 1999 Ministerial Meeting in Seattle and the 2001 Summit negotiations of the Free Trade Area of the Americas in Quebec City appear to illustrate the erosion of support for regulatory integration according to the traditional trade approach.

Despite the shortcomings of the traditional trade approach, the important problem of transatlantic regulatory regionalism facing GM crops remains. Is there a regulatory integration strategy that could overcome this regulatory regionalism while remaining sensitive to the regional differences in regulatory development? Such an ideal framework for regulatory development and integration essentially combines the stable, operational and rules-based characteristics of the North American regulatory approach with the socially responsive characteristics of the EU approach into the regulatory development and integration strategy. Amending traditional trade diplomacy so that it deals with social regulatory barriers along these lines would restore public confidence in the ability of the international trading regime to manage international integration.

This study essentially represents an initial attempt not only to identify the challenge of transatlantic regulatory regionalism and social regulatory barriers to trade diplomacy, but further to propose a regulatory development and integration strategy capable of overcoming this challenge. An interdisciplinary research approach is employed because identifying a jurisdiction's regulatory development and integration strategies involves a survey of economic, political, sociological and legal factors.

Given the combination of timely, controversial topics – GM crops and international trade – and the interdisciplinary approach, it is hoped that this study will be useful for a broad audience among not only academics, students and policy-makers, but also among those simply interested in the regulation and trade of GM agricultural crops.

As this book has been adapted from my doctoral thesis at the London School of Economics and Political Science, I would like to acknowledge the time and expertise of my adviser, Mr Stephen Woolcock, given to my thesis. I would also like to acknowledge the following institutions for the generous funding for my doctoral programme: the Canadian Wheat Board (Canadian Wheat Board Doctoral Fellowship); the Benjamin J. Sanderson Trust (Benjamin J. Sanderson Fellowship); and the Suntory and Toyota International Centres for Economics and Related Disciplines (Suntory and Toyota Research Studentship).

Special thanks must also go to the vibrant biotechnology community at the University of Saskatchewan – both colleagues and students – whose ideas, comments and criticisms all influenced this study in innumerable ways (except, of course, in any errors or omissions, which are entirely mine).

Most of all, I would like to thank Shannon, my everything.

G.E. Isaac
University of Saskatchewan
Saskatoon, Canada
June 2001

Part I
The Issues

The international trade of genetically modified (GM) agricultural crops has drawn attention to a controversial and complex challenge facing trade diplomacy – the challenge of regulatory regionalism created by social regulatory barriers to trade. In fact, this study argues that this challenge leaves trade diplomacy at a crucial crossroads. Applying the traditional approach of trade liberalization to the market-access barriers caused by regulatory regionalism will increasingly erode public support for the international trade regime and marginalize trade diplomacy as a viable force in international integration. Amending the traditional trade-liberalization approach will, instead, enhance the prospects that trade diplomacy can effectively deal with regulatory regionalism and promote international integration. Essentially, the objective of this study is to define the traditional trade-diplomacy approach, identify the development of GM-crop regulations in order to understand why the traditional trade-diplomacy approach is unable to address the challenge of regulatory regionalism associated with GM crops and, given the shortcomings, how that approach may be amended in order to enhance regulatory integration and market access.

It is useful to begin with the question: how did trade diplomacy get to this crossroads? Basically, it has arrived here because of its success. The objective of trade diplomacy is simple – to enhance market access for traded products (including goods, services and investments) by removing barriers to trade. Initially, the focus was on reducing or removing border measures, such as tariffs and quantitative restrictions, by establishing rules disciplining the use of these measures against foreign products. Trade rules outlining the acceptable and unacceptable use of border measures were negotiated and established at the international level and then the rules were imposed 'top-down' on signatory countries. As border measures have come under international discipline, domestic regulations have assumed greater importance as a source of trade barriers. Domestic regulations may become the basis for trade barriers against foreign products whose production, processing or composition contravenes domestic preferences and concerns embodied in the domestic regulations. When regulatory approaches differ between jurisdictions, regulatory regionalism is created, potentially hindering market access and generating trade tensions.

Given the success of trade diplomacy in disciplining border measures, the same top-down approach has been applied to domestic regulations. This can only work if domestic regulations are comparable to border measures and, of course, they are not. Border measures generally target foreign companies. Domestic regulations aim to ensure goals such as food safety and environmental protection, as well as to meet moral, ethical and religious preferences. While the top-down imposition of trade rules on border measures may be perceived to have very little impact on domestic citizens, imposing trade rules upon the ability of a jurisdiction to set regulations for food safety and environmental protection according to domestic concerns and preferences can have an enormous impact upon domestic citizens. In short, as border measures and domestic regulations are not the same, the same solution does not apply. Indeed, as trade diplomacy has disciplined the type of domestic regulations that nation-states may use, it has reached deep into areas of national competence and entered controversial territory. Both the 1999 Ministerial Meeting of the World Trade Organization in Seattle and the 2001 Summit negotiations of the Free Trade Area of the Americas, besieged by a vast coalition of interest groups, exemplify the controversy that arises when traditional trade diplomacy and domestic regulations interact. The accusation generally made by those critical of trade liberalization is that it represents an invasive economic strategy designed to strip nation-states of their regulatory-policy autonomy in order to serve the profit-maximization interests of faceless multinational corporations. Given such a sensationally negative view of trade diplomacy, it is not difficult to understand why regulatory regionalism and regulatory barriers represent a controversial and complex challenge to trade-liberalization efforts.

The time has come for a critical assessment of the problems associated with applying the traditional trade-diplomacy approach to the market-access problems of regulatory regionalism and regulatory barriers to trade. Understanding the nature of domestic regulations is a critical factor in understanding potential strategies for increasing regulatory integration and decreasing market-access barriers.

It appears that there is no better case-study of this trade-diplomacy challenge than the international trade of GM crops. The development and commercialization of GM crops have prompted the establishment of what may be called social regulations to ensure domestically preferred levels of food safety and environmental protection and to address broader social norms, such as moral, ethical and religious concerns about the use of advanced technologies. These social regulations are fundamentally shaped by endogenous political-economy factors that can be quite unique to a particular regulatory jurisdiction. In fact, differing political-economy factors within North America and the European Union have produced transatlantic regulatory regionalism, where GM crops approved as safe in North America and widely used in commercial production have been delayed and denied market access in the European Union because of social regulatory barriers. While Canada and the USA have proposed dealing with the trade tensions through the traditional trade-diplomacy approach of the World Trade Organization, the European Union has rejected this course of action. In other words, trade diplomacy is at a crucial crossroads, where a transatlantic stalemate has emerged with no apparent method in sight for dealing with the regulatory regionalism.

Therefore, this study is a critical assessment of the challenge of social regulatory barriers to trade diplomacy through a case-study of the transatlantic regulatory regionalism associated with GM crops. It will be argued that overcoming regulatory regionalism requires a strategy of regulatory integration, and yet the prospects and limits of regulatory integration are a function of the domestic regulatory development process shaped by endogenous political-economy factors. Hence, this must be a bottom-up critical assessment beginning with the political-economy factors shaping the development of domestic regulations in North America and in Europe, followed by an examination of the prospects and limits of transatlantic regulatory integration.

Such a bottom-up critical assessment of the regulatory development and integration strategies pursued within a regulatory jurisdiction requires an interdisciplinary research methodology examining economic, political, sociological and legal factors. Accordingly, an international political economy (IPE) approach is employed because of its two fundamental characteristics. First, the broad aim of an IPE approach is to analyse the structure of the relationship between the state and the market including its evolution, its dominant actors, its stability (or instability) and its future trajectory (Gilpin, 1987; Strange, 1991). Of course, as both states and markets function on many levels (i.e. local, regional, national, bilateral, plurilateral and multilateral) an IPE approach must simultaneously deal with the state–market relationship across these different levels. For instance, in one approach, Strange (1988) identifies four key factors to assess the structure of the state–market relationship – knowledge, finance, production and security – and uses these factors to understand the structure of the international economic system. Secondly, and flowing from the first, an IPE analysis of regulatory development and integration requires a simultaneous assessment of the political-economy factors – economic, political, sociological and legal dimensions – within a particular regulatory jurisdiction. Indeed, the integration of economics, political science, sociology, international law and international relations in Strange's (1988) key factors is obvious. Dealing with these factors simultaneously is contrary to conventional, discipline-specific research approaches. For instance, Malchup (1979) argues that economic approaches deliberately hold these other dimensions constant by assuming they are static, exogenous variables to the economic system. Yet, in analysing the development of a regulatory approach and a regulatory integration strategy within a particular jurisdiction, it would be unacceptable to focus on one dimension and assume that the rest have no impact on the structure of the regulatory approach. The benefit of the IPE focus on the interaction between these dimensions is that it avoids the 'trap of institutionalism', whereby analysis from one perspective in isolation fails to recognize the belief system that the perspective is based on and which is central to the subsequent analysis (Dezalay, 1996). Accounting for the belief systems underlying the other dimensions is to adopt a 'reflexive methodology' avoiding the trap of institutionalism (Bourdieu and Wacquant, 1992).

This case-study fits these two fundamental characteristics of an IPE approach. The analysis of social regulatory barriers, regulatory regionalism and regulatory integration is essentially an analysis of the structure of the transatlantic state–market relationship. This requires analysis of the 'regulatory arena' (Dezalay, 1996) that exists between the state and the market (Strange, 1988), both within a regulatory jurisdiction, as the domestic regulatory approach is developed, and between regulatory jurisdictions, as regulatory integration occurs. Further, there are significant limitations in assessing this research problem from one dimension only. For instance, with respect to the economic dimension, there is a lack of quantitative analytical options. Social regulatory barriers are not quantitative trade-restriction measures and the economic analysis of regulatory barriers is currently too complex and underdeveloped for systematic economic analysis (Roberts *et al.*, 1999). From a social-legitimacy point of view, the predominant calls for social regulatory barriers are from non-trade, social interests and not from traditional trade or commercial interests (Perdikis *et al.*, 1999). These non-trade interests view 'social protectionism' as a legitimate constraint on trade liberalization and they reject attempts to quantify social regulatory barriers as economic costs that create market failures. Instead, they support the notion that regulations play a crucial social function that cannot be decomposed simply into an economic cost. Therefore, the IPE approach offers an appropriate methodology for examining the regulatory regionalism that is having an impact on the transatlantic trade of GM crops.

A particularly important methodological issue is the time frame of this study, which includes transatlantic regulatory development and integration issues up to June 2001, with the major focus on those issues that emerged in the late 1990s. Essentially this means that this study is not a critical assessment of a closed issue. Instead, it is a critical assessment of a complex and controversial open trade-diplomacy challenge that is expected to become the most crucial issue on the trade agenda.

Given the controversy surrounding the development, commercialization and international trade of GM crops, it is important, at this point, to specify the scope of this study. This study is not a critical assessment of the science of genetic modification. Rather, it is a critical assessment of the social implications of GM crops – specifically, the way that this science is promoted and regulated in North America and in Europe and how these approaches are in concert or conflict with the international trade regime. This study adopts four premisses. First, the focus of agricultural production must be on achieving greater environmental sustainability while meeting demands for safer, higher-quality products. Secondly, as the dominant agricultural production system is the intensive agricultural system, it is most practical to examine ways to make it more environmentally sustainable. Thirdly, and perhaps most controversially, GM crops can be congruent with a sustainably intensive (Sage, 1999) agricultural production system, using, for instance, fewer synthetic chemicals to achieve greater productivity while meeting consumer demands for safer and higher-quality products. This does not imply, however, that this study unreservedly supports technological progress over technological precaution. Instead, the fourth premiss of this study holds that it is the application, management and distribution of technology that set the parameters for technological precaution and the boundaries for technological progress.

Consistent with the research methodology and the four premisses identified above, this study is organized into four parts. In Part I, the foundation of the study is laid. In Chapter 1, concepts such as social regulatory barriers to trade and regulatory regionalism are more precisely defined. In Chapter 2, there is a brief examination of the science of agricultural biotechnology, its application to crop production and the economic implications of this development for the agricultural sector, as well as an examination of the consumer concerns that have created so much controversy. Together, Chapters 1 and 2 establish the context for analysing regulatory regionalism created by social regulatory barriers.

In Part II, the challenges of regulatory development are linked with the challenges of regulatory integration. In Chapter 3 there is a discussion of those economic interests supporting the traditional trade-diplomacy approach to regulatory development and integration. This includes a comprehensive assessment of how the international trade regime deals with social regulatory barriers. In Chapter 4, there is a discussion of the regulatory development and integration approaches supported by social or non-economic interests opposed to the traditional trade approach. In Chapter 5, the competing interests are brought together to examine the vociferous debates associated with the development and integration of GM-crop regulations. Seven framework regulatory debates associated with regulating technology and five specific debates associated with regulating GM crops are identified. It is shown that the regulatory integration strategy pursued by a jurisdiction actually flows from the specific GM-crop regulations adopted, which in turn flow from the broader framework for regulating technology within the jurisdiction. In short, Part II establishes a conceptual model for determining the regulatory framework, the type of specific regulations and, consequently, the regulatory integration approach possible within a particular jurisdiction.

In Part III, there is a case-study of transatlantic regulatory regionalism. In Chapters 6 and 7, respectively, the North American and the European Union approaches to regulatory development and integration are identified, according to the conceptual model developed in Part II.

In Part IV, Chapter 8 brings together Parts I, II and III in order to assess why the traditional trade-diplomacy approach is incapable of addressing the transatlantic regulatory regionalism associated with GM crops and, hence, stands at a crucial crossroads. Given this analysis, an amended trade-diplomacy approach is proposed and examined.

1 Social Regulatory Barriers

The objective of this chapter is to define social regulations, identify their characteristics, examine how they may create regulatory barriers and then discuss how they challenge international trade. Accordingly, this chapter is organized as follows. First, social regulations are defined, along with their propensity for instability and their trade-barrier aspects. Then international trade is discussed within the broader concept of international integration. Next a theoretical framework tying the development of regulations within a jurisdiction to the prospects and limits of regulatory integration between jurisdictions is proposed. The chapter concludes with an introduction to the specific case-study – the transatlantic trade of genetically modified (GM) crops.

1.1 Regulations: Instability and Market-access Barriers

When discussing regulations, it is important to disentangle specific regulations from the broader regulatory framework. The regulatory framework includes the meta-principles and features of the regulatory approach within a jurisdiction such as: regulatory accountability to an elected official or to an arm's-length, independent agency; regulatory principles the regulators must achieve (e.g. safety/hazard prevention only or broader socio-economic development targets); and the type of liability laws that exist. These principles essentially define the parameters within which specific regulations are established.

Regulations are the specific rules – developed within the regulatory framework – that are mandatory, legal measures adopted by governments to influence the outcome of economic and social activities within the regulatory jurisdiction, which is often synonymous with the nation-state. Caswell and Henson (1997) argue that regulations can be in the form of either direct regulations or liability laws. The former are proactive guidelines for acceptable behaviour while the latter are reactive measures used to discipline behaviour found to be unacceptable. Together, direct regulations and liability laws establish a relevant rule of law within the regulatory jurisdiction.

In a very broad sense, regulations perform two functions. The first is an economic function, such as ensuring the productivity and competitiveness of commercial activity within a particular jurisdiction (Porter, 1990). This can include tax measures to stimulate investment in research and development, competition laws to ensure the presence of a competitive commercial environment and employment programmes to support worker training and skills acquisition strategically targeted to the jurisdiction's competitive needs. Essentially, the economic function of regulations is to improve the allocative efficiency of the market system. The second general function of regulations is a social function, which channels market activity to meet non-economic preferences and expectations within a jurisdiction. Social regulations include, for example, food-safety measures (Spriggs and Isaac, 2001), environmental-protection measures (Wheale and Williams, 1993) and labour standards (Langille, 1996), as well as measures to address social norms, such as moral, ethical and religious concerns. In contrast to the economic function, the social function of regulations tends to focus on improving the equity – the distribution of benefits and burdens – within the jurisdiction. Of course, this categorization of regulations is simplistic, because all regulations are, in fact, a blend of economic and social objectives. Yet the purpose of this categorization is to illustrate that some regulations are dominated by efficiency or competitiveness concerns (economic regulations), while other regulations are dominated by equity or non-commercial concerns (social regulations). Furthermore, as will be discussed below, this categorization of regulatory function extends to a categorization of various stakeholders according to the type of regulatory development and integration strategies they support.

The development of a regulatory framework and specific regulations within that framework involves building consensus and cohesion through a political process of balancing the rights and interests of stakeholders within the regulatory jurisdiction (Dezalay, 1996; Picciotto, 1996). Given the political balance required, once the framework is set, institutional stability is necessary for the rule of law to become established in a manner beneficial to both the economic and social functions. For instance, economic benefits of a stable regulatory framework include commercial certainty and predictability, while social benefits include confidence in the regulatory approach. Indeed, a regulatory framework that changes frequently appears to indicate that regulators lack control, thus decreasing public confidence in the regulatory approach. In short, economic and social benefits arise when a regulatory framework is stable and operational.

Accordingly, changing the regulatory framework is not an easy process. It has been argued that institutional inertia makes change difficult, requiring a high-level political push (Frost, 1998). The impetus for such a political push is often found in changes in the economic and social dimensions of the regulatory jurisdiction. Two crucial causes are technological innovations and political crises (Dezalay, 1996). These causes are quite relevant to this study and there exists a synergistic relationship between them. The public may have a poor understanding of advanced technologies, and their development and commercialization can cause a political crisis as public concern grows about the level of regulatory oversight governing the risks of the new technology. Further, the introduction of advanced technology can set the economic function of regulations in conflict with the social function. For instance, the innovation may promise economic productivity and efficiency benefits, so that economic regulations aim to encourage technological progress. In this case, economic regulations are used to set the pace and dispersion of advanced technology (Zilberman *et al.*, 2000). On the other hand, public anxiety about the innovation requires social regulations encouraging technological precaution in order to maintain public confidence in the regulatory approach (Kraus, 1998). Maintaining stability in the regulatory framework can be extremely difficult in the face of such political pressures, jeopardizing both the economic and the social benefits that flow from regulatory stability.

While stability may be an appropriate goal of the regulatory framework, it is not necessarily the appropriate goal of the specific regulations. For instance, given significant public anxiety about the commercialization of advanced technologies, flexibility in specific regulations may be necessary so that the regulations are seen to be reacting to the anxiety in a publicly acceptable manner. Stability, in this case, may appear to result in regulatory decisions about product approvals that fail to account for public concerns. Therefore, stability in the regulatory framework is necessary, but the specific regulations may need to be flexible in order to be publicly acceptable.

1.1.1 Regulatory instability: advanced technologies

As mentioned above, the development and commercialization of advanced technologies, such as GM crops, is rife with controversial debates that motivate political challenges to both the specific regulations and the underlying regulatory framework, resulting in regulatory instability as the framework and the subsequent specific regulations are revised. For instance, developers of GM crops have argued that regulations are not focused enough on technological progress, while critics have argued the opposite – that regulations are not focused enough on technological precaution. In fact, as will be discussed in more detail below, supporters tend to emphasize the economic function of regulations in achieving technological progress, while critics tend to emphasize the social function in achieving technological precaution. The controversial debate is over the appropriate function of regulations; should they be to maximize technological progress or technological precaution?

The common approach to regulating new technology in North America and in the European Union is according to the risk-analysis framework (RAF). First outlined in 1983, the RAF seeks to identify procedures to effectively perform risk assessment, risk management and risk communication, where the objective is to establish stable and predictable regulatory principles upon which to base the specific regulations (National Academy of Sciences, 1983).

Regulatory instability arises when there are significant debates associated with how the RAF should be applied. Such debates generally reflect conflicts between those who seek to maximize technological progress and those who seek to maximize technological precaution, making it extremely difficult to establish a cohesive and consensual regulatory framework. While the RAF debates associated with GM-crop regulations will be examined in greater detail in Chapter 5, it is important to emphasize that stable framework principles for regulating GM crops have not been established within regulatory jurisdictions, let alone the specific regulations within this framework. Supporters and critics of the technology are actively lobbying government regulators to establish regulations congruent with their views on technological progress and precaution. In reaction, government regulators have been unable and perhaps unwilling to nail down the framework principles for the regulatory RAF. It will be shown (in Part III) that, even in jurisdictions with seemingly established and stable regulatory frameworks, there is evidence of regulatory instability.

1.1.2 Social regulatory barriers to trade

Unstable regulations can incur market-access barriers during the international trade of advanced-technology products, because political-economy factors, such as political processes and the influence of various economic and social interests, differ between jurisdictions. Consequently, divergent jurisdictions develop different regulatory frameworks and specific regulations can differ significantly. Hence, divergent regulations can become the basis for market-access difficulties faced by foreign products whose production, processing and/or composition contravene the preferences and expectations embedded in the regulatory framework and specific regulations of the domestic market, despite the fact that these products have been approved in their home market.

Up to this point, economic and social regulations have both been examined. Yet, in the context of regulatory barriers to trade, a distinction between the two must be made. Social regulatory barriers represent a new and complex challenge for trade diplomacy. Table 1.1 gives a categorization of various types of trade barriers under two broad headings: tariff barriers and non-tariff barriers to trade. Trade diplomacy has been successful in reducing tariffs on manufactured goods and in bringing greater discipline on the use of non-tariff barriers, such as quantitative restrictions (Grimwade, 1996; Jovanovic, 1998e).

Yet trade diplomacy has not been as successful with regulatory-type barriers to trade. Building on the general distinction introduced above between the economic and social function of regulations, there are two types of non-tariff regulatory barriers to consider. Economic regulatory barriers tend to be driven by commercial interests, who seek government assistance in competing with foreign firms. In this sense, economic regulatory barriers represent economic or commercial protectionism, which is often deliberate and can take the form of, for instance, commercial laws on foreign direct investment and ownership. Alternatively, social regulatory barriers tend to be driven by non-commercial interests, such as consumer and environmental organizations. In this sense, social regulatory barriers represent social protectionism. They differ from economic regulatory barriers because often the trade-barrier aspect of social regulations is a secondary consequence; that is, social regulations are established to meet domestic social preferences and expectations, although they may also have an impact upon trade flows.

Table 1.1. Classification of trade barriers.

	Non-tariff barriers		
		Regulatory barriers	
Tariff barriers	Quantitative restrictions	Economic	Social
Tariffs on imported products	Import quotas Voluntary export restraints Anti-dumping duties Countervailing duties	Taxation laws Competition laws Foreign investment laws	Food safety Environmental protection Labour standards Cultural/normative protection

While regulatory barriers in general and some social regulatory barriers in particular are not new to the trade agenda, the depth of food-safety and environmental-protection regulations within national regulatory competence is a complex challenge for trade diplomacy. For instance, the 1995 Agreement on Sanitary and Phytosanitary Standards (SPS Agreement) under the World Trade Organization (WTO) was explicitly aimed at disciplining the use of food-safety-type social regulatory barriers (see Chapter 3). Rather than facilitating an international harmonization or centralization of domestic food-safety regulations, the SPS Agreement has instead revealed that social regulations are a formidable challenge to traditional trade diplomacy. In a controversial trade-dispute decision, the WTO ruled that a European Union (EU) ban on imports of beef treated with growth hormones violated the SPS Agreement, and yet the EU has not abided by this decision and has not modified its domestic regulations to be congruent with the SPS Agreement. In defence of their social regulatory barriers, the EU is willing to remain in contravention of the WTO trade regime and this should stand as a vital warning to trade-diplomacy efforts at disciplining social regulatory barriers.

Before moving on to the discussion of international integration, it is important to clarify the concepts of social regulatory barriers and regulatory regionalism. Social regulatory barriers are market-access barriers facing imported products because they fail to meet domestic social regulations. Regulatory regionalism exists when not only do regulatory jurisdictions differ on specific regulations but they also differ on fundamental framework regulatory principles. For instance, there is no internationally harmonized regulatory framework for GM crops. Instead, there are two dominant regulatory jurisdictions – North America and the EU – with different regulatory frameworks and specific regulations. This transatlantic regulatory regionalism has created a trade situation where GM crops approved as safe for food and feed use and for the environment in North America have faced delayed market-access approvals according to EU social regulations. The key point to make is that, while the traditional trade-diplomacy approach may be able to deal with differences in specific social regulations when the underlying frameworks are congruent, it appears that there are significant challenges in dealing with regulatory regionalism that results from differences in underlying regulatory frameworks.

1.2 International Integration

As the thesis of this study is that traditional trade diplomacy cannot adequately deal with the challenge of transatlantic regulatory regionalism associated with social regulatory barriers to GM crops, the assessment must now identify the traditional trade-diplomacy approach, along with competing or alternative approaches. The most illustrative way of accomplishing this is within the broader context of international integration.

International integration is the dynamic process whereby the economic and social (i.e. political, cultural, normative, etc.) dimensions of a nation converge with those of other nations. International integration occurs regionally (bilaterally or plurilaterally) and globally (multilaterally). Moreover, it is a neutral term. For supporters, on the one hand, international integration is predicated on the belief that collective action among individual nation-states can lead to greater overall gains than those possible when nations act alone. This belief is best exemplified by the mandate of the United Nations, which may be summarized as: to achieve global stability and security by promoting economic and social development through the coordinated efforts of individual nation-states. On the other hand, critics view international integration as a corrosive force that erodes national distinctiveness. This study does not aim to address the debate over the efficacy of integration. Instead, it adopts the premiss that, good or bad, integration is inevitable and the important issue is not about choosing between integration and disintegration but about managing integration in an acceptable manner.

The challenge of social regulatory barriers to international trade diplomacy is in essence a challenge for international integration, because trade is really only one aspect of international integration, although a highly visible aspect. As the commercial activities of independent nation-states have become increasingly enmeshed, differences in social regulations have become more apparent. These differences have hindered integration and, in some cases, have led to calls for greater social protectionism against international integration.

A notable characteristic of the literature on international integration is the separation of international economic integration from international social integration. International economic integration adopts economic principles to explain appropriate regulatory development and regulatory integration strategies for a jurisdiction. Accordingly, rational, optimizing behaviour fulfilling the economic function governs policy development, shapes social regulations and forms the basis for dealing with social regulatory barriers through trade diplomacy. In fact, it will be argued that this perspective has been the basis for the traditional trade-diplomacy approach. Alternatively, international social integration adopts a perspective that argues that markets are embedded in normative constructs so that the economic perspective is meaningless if separated from social realities. It follows that regulatory development and integration strategies must target more than just an economic function; they must also fulfil important social functions. As a result, the social perspective suggests that trade diplomacy must be more socially responsive when dealing with social regulatory barriers.

Crucially, the economic and the social perspectives on international integration support divergent paths for regulatory development within a jurisdiction and divergent recommendations on how regulations should be integrated between jurisdictions. The key to understanding the challenge of social regulatory barriers to traditional trade diplomacy is to understand the differences in the two perspectives, which are discussed in greater detail below.

1.2.1 Economic perspective: scientific rationality

Since at least 1948, the economic perspective has dominated trade diplomacy, which has been a relatively successful example of international economic integration. It has facilitated the development of a rules-based international trade regime characterized by commercial regularity, orderliness and predictability, allowing for a substantial growth in global trade flows (World Trade Organization, 1995).

The basis of the economic perspective is that decentralized and distributed decision-making can lead to economically optimal outcomes (Bratton *et al.*, 1996). As the key goal is to improve market efficiency and effectiveness, there is a deliberate separation of economic dimensions from social dimensions in order to depoliticize the regulatory-development process. This removes protected positions that create social and political market failures, hindering the attainment of economically optimal outcomes. Removing social and political market failures permits a focus on real market imperfections with first-best policies.

This does not imply, however, that social dimensions are sacrificed. Instead, it is held that improved market operations increase growth and development. This, in turn, increases income and the demand for social regulations, such as food safety and environmental protection, because, it is argued, they are income-elastic; as incomes rise, so does the demand for these types of regulations (Wang and Caswell, 1997). These income-elastic social preferences are internalized within the competitive strategies of private firms for three reasons: consumers demand them; they enhance the firm's social reputation; and they increase investor confidence (Woolcock, 1996). In short, the economic perspective holds that it is vital to first remove social and political market failures in order to produce the best conditions for economic growth and that this in turn allows for the establishment of optimal social regulations, with a tendency to become more stringent, not less – a regulatory race to the top.

The economic perspective has been used to assess both how regulatory development should occur within a regulatory jurisdiction and how regulations should be integrated between jurisdictions. In respect of how regulatory development should occur within a jurisdiction, public policies in the postwar period were significantly influenced by Keynesian economics and its reliance upon government policy expertise. It suggested that government intervention in the market could secure an optimal outcome that the decentralized market was unlikely to produce. A common argument, for instance, is that government intervention preserves a long-term agenda to facilitate the economic prosperity of the nation while the operation of decentralized market forces tends to focus only on short-term issues. Underlying this theory were two important assumptions. First, it was assumed that policy-makers are aware of all economic and social preferences through the political process and could design first-best regulatory policies, producing optimal outcomes. Secondly, it was assumed that government policy-makers are benevolent agents whose primary goal is the improvement of national welfare and that no other objectives obfuscate this goal. These two assumptions did not go unchallenged.

Public-choice theory challenged the ability of policy-makers to be aware of all economic and social preferences. Samuelson (1954) argued that the political process could not accurately reflect all preferences because the nature of public goods encourages market failure. On the one hand, voters can under-represent their utility derived from a public good because they have not had cause to consume the good (Samuelson, 1954). An example might be local fire services; someone who has never experienced a house fire might undervalue the local fire service. Alternatively, voters can over-represent their utility because they do not know the true cost of public goods such as food-safety or environmental-protection measures. Malchup (1979) argued, for example, that, if the true cost were known, then the demand for the public goods would fall.

Similarly, social-choice theory also challenged the assumption that government policy-makers could be aware of all economic and social preferences. Arrow (1963) argued that, because voting paradoxes prevent the aggregation of all preferences, the preferences that policy-makers target are, at best, only partial preferences. Further, Olson (1965) argued that the problem does not just rest with the voting process, because regulators are easily captured by and respond to special interests, resulting in government intervention in pursuit of only limited objectives, not overall national welfare.

Both public- and social-choice theory argued that, despite the best efforts of government policy-makers, the combination of failures with the political process and the prevalence of regulatory capture could easily result in protected political positions. Without an awareness of all economic and social preferences, a true first-best regulatory policy could not be established. In order to correct these policy problems, the public- and social-choice theories adopted the economic perspective of depoliticizing the regulatory-development process, by separating the economic objectives of market efficiency from social objectives of equity. Once the market was working efficiently, then optimal levels of social regulations could be established, rather than the reverse process, where suboptimal social regulations are first established, which then hinder market efficiency. Additionally, it was proposed that a subsidiarity or devolution policy be adopted in the regulatory development process in order to decrease the variance in economic and social preferences, decrease the number of special interests lobbying the process and enhance the link between the regulators and those that the regulations affect. The economic perspective thus supports a depoliticization and a competitive decentralization of the regulatory development process.

The economic perspective also led to challenges to the second assumption of the Keynesian-type interventionist theories – that government policy-makers are benevolent maximizers of national welfare. Clearly, government does play a more active role, including intervening to fulfil its own preferences and objectives. For instance, according to the constitutional political-economy approach, Brennan and Buchanan (1980) cautioned against assuming that government regulation is in pursuit of social welfare. They argued instead that governments are 'regulatory monopolists'. At the governmental level, the political concerns of re-election dominate long-term economic and social development concerns. Within government, departments compete with one another to secure revenues and preserve regulatory power or authority. In this sense, regulations must be viewed, at least partially, as political instruments created by policy-makers in pursuit of their own self-interest and not in pursuit of national welfare. To guard against such possible misuse of regulatory power, it was argued that the economic perspective be adopted to prevent the political market failures caused by the self-interest of policy-makers.

In respect of the political-crisis driver for regulatory change, the economic perspective has little support for a reactive, socially responsive regulatory approach. Instead, it supports a regulatory framework that is depoliticized in order to disentangle political and social market failures from real market failures (Majone, 1990). Underlying this perspective is the belief that government regulatory intervention has an adverse impact on economic and social prosperity, because it increases market distortions, driving up costs and decreasing commercial competitiveness, productivity and market efficiency (Woolcock, 1998). This, in turn, decreases economic growth and development and decreases the demand for income-elastic social regulations.

In respect of the technological-innovation driver for regulatory development and change, the economic perspective generally assumes that technology and innovation are vital factors of economic growth and welfare.[1] As a result, it supports a regulatory framework that encourages technological progress. For instance, it is quite common for economic analysis to support 'scientific-rationality' approaches to regulating the risk of new technology (van den Daele *et al.*, 1997). The economic- and scientific-rationality perspectives are similar, in that they decompose complex behaviour and actions into causal–consequence models, which are then used to forecast outcomes (this is discussed further in Chapter 5).

In short, the economic perspective on regulatory development holds that an efficient regulation is one that corrects a market failure and improves the resource allocations of the free market, encouraging technological innovation and growth. Accordingly, policies must be activated by market failure and focused only on fulfilling economic functions in order to improve the prospects for economic growth, allowing for the optimal establishment of social regulations, free from the distortions of protected political positions.

International economic integration is simply an extension of this perspective on regulatory development to the international level, where the nation-state, as the regulatory jurisdiction, remains the primary actor. The rationale for international economic integration is to facilitate scale economies and the optimal allocation of factors of production according to the principle of comparative advantage (Jovanovic, 1998e). Comparative advantage is simply the principle that global welfare can be maximized by allocating factors of production to where they are most efficiently utilized and then trading products to meet consumer demand. A division of the factors of production is widely accepted as an organizing principle at the national level, and comparative advantage simply extends this notion to the international level (Malchup, 1979).

According to the international economics literature, trade barriers hinder market access and distort comparative advantage. They fragment international markets, limit economies of scale, increase costs and create market-access uncertainty. In fact, economic trade analysis considers barriers to be 'economic costs' placing distortionary constraints on a firm's production function and, collectively, on a nation's production-possibilities frontier. Such an analysis focuses on the cost, price and allocation effects of trade barriers in order to demonstrate their impact on comparative advantage and economic welfare. In fact, economic trade analysis has long shown comparative advantage to be a compelling argument for non-discriminatory, liberalized trade, because it results in the optimal and efficient allocation of production

[1]It is important to note that, while this is a widely established economic tenet, not all economists hold this view. For an insightful discussion of the 'steady-state' or 'zero-growth' economy, see Daly (1971), Daly and Cobb (1994).

and consumption patterns. Indeed, in the literature, there is general acceptance that any type of economic integration is positive (Grimwade, 1996). However, debates exist regarding the level of economic integration that is most optimal in terms of national welfare. For instance, Jovanovic (1998a–e) assesses the economic benefits of and limitations to greater economic integration at a regional (bilateral or plurilateral) and global (multilateral) level.

As mentioned above, international economic integration has been the foundation of trade diplomacy since at least the establishment of the General Agreement on Tariffs and Trade (GATT) in 1948. Trade diplomacy aims to separate economic integration from social integration in order to disentangle political and social market failures from real market failures. According to the WTO, the goal of trade diplomacy – based on the international economic-integration perspective – is to develop multilateral rules to remove social and political protectionism – to depoliticize trade and make it a function of comparative advantage, not political advantage (WTO, 1995). The threat is that political and social market failures would become locked into the regulatory approach, preventing the identification of an optimal regulatory standard and hindering economic integration, innovation and growth (Katz and Shapiro, 1986; David, 1987; Arthur, 1989).

Further, on the contentious issues of integration and sovereignty, the economic perspective holds that economic integration does not imply a loss of policy sovereignty or autonomy. It is argued that international trade treaties or agreements represent concessions by all signatories in order to realize mutually beneficial gains from economic integration (Jovanovic, 1998e). There is no real loss of policy sovereignty as all signatories or contracting parties have given up the exact same degree of autonomy. For example, in discussing the impact of the Canada–US Free Trade Agreement on Canadian sovereignty, Lipsey (1988) concluded, 'Canada can establish and maintain its own distinctive social policies, while liberalizing its trading arrangements with other countries.' In other words, economists tend to argue that economic integration and trade liberalization do not result in an absolute loss of domestic policy autonomy *vis-à-vis* other countries.

It should be noted, however, that the discussion on the economic perspective is not to suggest that social dimensions have never been a factor in the economic analysis of integration. Indeed, both Cooper and Massell (1965) and Johnson (1965) introduced a 'public-good' rationale for regional economic integration, whereby regional as opposed to multilateral economic integration permitted the increased production and consumption of public goods. Also, agricultural economics has often adopted a view of economic integration where social dimensions have played a primary role in trade-policy development. For instance, popular social arguments for supporting and protecting domestic agriculture include strategic arguments of ensuring a domestic food supply and 'multifunctionality' arguments claiming that the agriculture sector not only produces food but that it also provides social benefits in the form of the protection and preservation of the countryside and the rural lifestyle. Political-economy models of tariff determination have been developed to cope with the agricultural sector's social protectionism (Frey, 1984, 1985). The key point is that the analytical focus is on determining the effects of the social dimensions on economic efficiency; hence the economic perspective, not the social perspective, dominates.

There are two fundamental trade concerns about the impact of social regulatory barriers and regulatory regionalism upon international economic integration. First, social regulatory barriers are seen as an easy target for protectionist rent-seekers, because there is virtually no discipline on their use under international trade rules. Secondly, income-elastic social regulatory barriers could fragment international markets into exclusionist social regulatory jurisdictions that hinder multilateral economic integration and consequently limit the gains from trade. To deal with these concerns the economic perspective advocates that social regulatory barriers be subject to multilateral economic integration rules according to the traditional trade-diplomacy framework of the WTO. That is, social regulatory barriers should be subject to an economic interpretation of efficiency, they should have a minimum trade impact and trade diplomacy should avoid considering social arguments justifying the trade barriers (the integration strategy of the WTO will be discussed in more detail in Chapter 3).

This regulatory integration approach supported by the economic perspective is one of regulatory competition associated with the 'free-traders' school of trade theory (Bhagwati and Hudec, 1996; van Scherpenberg and Thiel, 1998). The regulatory-competition paradigm rejects efforts to coordinate or harmonize national regulations first, because it is argued that regulatory differences are considered to be a basis for comparative advantage and commercial opportunities (Lavoie and Sheldon, 1999). For instance, while income-elastic social regulations, such as food safety and environmental protection, would be different among trading partners at different levels of economic development, this may create a comparative advantage and gains from trade for firms in jurisdictions with a different regulatory framework. In other words, this paradigm supports national sovereignty or authority over regulatory policy, because of the belief that distributed regulatory development enhances both commercial opportunity and social welfare. Accordingly, it favours shallow international integration, with no aim to develop international social and political institutions other than those necessary to facilitate a rules-based regulatory framework for international economic integration.

Over time, the international regulatory-competition approach may produce regulatory integration, where either the market decides which approaches are optimal or a trade-dispute mechanism determines which approaches are permissible from the perspective of a rules-based framework. Alternatively, the regulatory-competition paradigm may not produce regulatory integration at all, such that regulatory divergence would prevail.

Therefore, in conclusion, the economic perspective suggests that regulatory development should focus on technological progress and market efficiency, while regulatory integration should focus on economic integration according to the regulatory-competition paradigm. The result is support for a scientifically rational, rules-based regulatory framework ensuring stability, orderliness and predictability during the process of international trade. With respect to the international trade of GM crops, this would take the form of an international regulatory framework focused on technological progress and market access, with very little regard for social dimensions in the regulatory process beyond fundamental principles such as safety or hazard. In fact, as will be argued in Chapter 3, this is precisely the approach adopted and promoted by the WTO.

1.2.2 Social perspective: social rationality

An alternative explanation of regulatory development and integration may be characterized as the social perspective or as social rationality. Fundamentally, the social perspective reverses the causation between market performance and social objectives by insisting that underlying the 'market' is a normative construct composed of domestic preferences, concerns and expectations (Bratton *et al.*, 1996). This normative social construct, including moral, ethical and religious concerns, cannot be separated from the market, because market operations occur within this construct. Consequently, the social perspective holds that neoclassical economic policies do not represent an independent ideal but rather a set of policies congruent with a neoliberal normative social construct. Without this construct, these economic policies would not be appropriate. Hence, the social perspective posits that economic policies are crucially constrained by the particular normative construct that exists within a jurisdiction such that understanding the regulatory development and integration strategies supported within a jurisdiction requires an understanding of this construct.

According to the social perspective, the development of a regulatory framework within a nation-state is not simply an exercise in correcting market failures. Instead, regulatory development also plays a more important social function in channelling economic activities to achieving normative objectives identified through the domestic political process. In this sense, social responsiveness is the main driver for regulatory change where regulators are expected to react to social dimensions, such as political crises, in an accountable manner. Moreover, regulatory effectiveness is assessed according to whether or not regulations successfully respond to legitimate social concerns, such as equity, as opposed to whether or not they meet economic measurements of efficiency. In contrast to the economic perspective, underlying the social perspective is the belief that government regulatory intervention improves economic and social prosperity, because it increases public participation and accountability, improving productivity and competitiveness (Woolcock, 1998). Indeed, from this perspective, it has been argued that the economic perspective leads to a 'hollowing out' of the nation-state, whereby social dimensions are sacrificed in order to enhance economic performance (Picciotto, 1996).

In respect of the technological-innovation driver for regulatory development and change, a popular sociopolitical treatment of the social perspective may be found in risk-society theory (Beck, 1992) and its extensions (i.e. Grove-White *et al.*, 1997), which focus on the regulation of risks from new technology. This 'social-rationality' approach holds that it is insufficient to view new technology and innovations simply as a positive force in economic growth. Instead, the social implications of science must be considered and, under this consideration, new technology may not always be greeted without reservation – despite its potential to improve economic growth. For instance, Habermas (1971) cautioned that the scientific complexity of advanced technologies erodes the established political process, because it has rendered elected decision-makers nothing more than 'mere agents of a scientific intelligentsia', where policy decisions are made about new technologies outside democratic accountability and in pursuit of economic objectives only. The social-rationality approach argues that it is the social dimensions, such as the application, management and distribution of the technology that are most crucial in deciding whether or not a new technology is necessary. Hence, the general tendency is to support regulations that encourage technological precaution and which are capable of responding to broader social concerns about new technology beyond just the potential economic benefits (Beck, 1992). In short, the social-rationality view is that technology brings change, but change disrupts the normative framework, such that precaution must be exercised before embracing technological change. Contrasting the two perspectives, then, the social perspective supports social rationality and technological precaution over the scientific rationality and technological progress supported by the economic perspective.

With respect to the international integration of regulations, a key concern of (but not limited to) the social perspective lies with the impact of integration upon the policy autonomy of the nation-state – as the primary regulatory jurisdiction – to support its own unique normative social construct. Does international integration strengthen or weaken the sovereignty and authority of the nation-state? With respect to the former, it has been argued that increased integration is not necessarily an internationalization of the nation-state, but rather a domestication of the international arena – strengthening the social authority of the nation-state (Hanreider, 1978, in Picciotto, 1996). This is possible when integration is based on a shared normative social framework (Milward, 1992), possibly promoting a regional or even global normative framework (Global Governance, 1995). On the other hand, it has been argued that the social authority of the regulatory jurisdiction is eroded, as nation-states accept limits on their power (Dezalay, 1996; Picciotto, 1996) and adopt an increasingly top-down framework, reflecting the economic perspective, unaccountable to the domestic political process and insensitive to the unique political-economy factors of particular regulatory jurisdictions (Giddens, 1994).

Given the ambiguous impact of integration on the policy autonomy of the nation-state, it is important to assess the social rationales for integration: that is, why would international social integration be supported at all? A fundamental rationale is the 'neofunctional' rationale, which posits that integration promotes interdependence, facilitating broader global stability and security objectives (Frost, 1998). To achieve this goal, integration is promoted by policy networks (Keck and Sikkink, 1994), epistemic communities (Haas *et al.*, 1992) and an emerging global civil society (Lipschutz, 1992; Shamsie, 2000).

Another rationale to pursue deeper social integration is to address social-externality issues or 'regulatory universals' (Dezalay, 1996), such as food safety and environmental protection. Social-externality issues are those whose impact is felt beyond the borders of the nation-state. Food products are widely traded and it is common for a consumption bundle to include foreign-produced goods. Similarly, environmental degradation ignores political boundaries and often has international consequences. Social integration is proposed as a method for ensuring that the externalities created in a foreign jurisdiction but which have an impact on the domestic jurisdiction are dealt with, rather than left to the forces of international economic integration. This can include methods to enhance the coordination of divergent food-safety and environmental-protection regulations among jurisdictions.

An important aspect of the social perspective is that it generally supports either a deeper regulatory integration between nation-states than that supported by the international economic-integration approach or no integration at all. This support is rooted in an increasingly popular dichotomy suggesting that a nation-state can pursue either greater international economic integration through liberalized trading arrangements or it can pursue stringent domestic social regulations, but not both (Garvey, 1995) – that is, increased trade liberalization is a negative, corrosive force that imposes market competition upon fundamentally non-market, social dimensions. Further, it is argued that, because non-market, social dimensions are improperly valued in the economic system, international economic integration leads to a social regulatory race to the bottom (Drache, 1996).

Unlike the economic perspective, the social perspective does not consider the international economic-integration approach of trade diplomacy to be depoliticized. On the contrary, it is argued that economic integration is, in fact, a highly interventionist form of political governance, imposing neoliberal economic policies on national governments for the benefit of only a minority of international capitalist entrepreneurs (Dezalay, 1996). Langille (1996) argues that it puts social polices beyond the authority of domestic governments. At various times, the international trade regime has been characterized as subversive liberalism (Rhodes, 1991, 1994), symbolic imperialism (Dezalay, 1996) and global unilateralism (Whitman, 1984; Strange, 1986) of the neoliberal economic perspective.

The social perspective's rejection of international economic integration alone creates a propensity to support social protectionism – regulatory regionalism based on social values and concerns, where jurisdictions cluster together to fend off the influence of international economic integration. Two general implications of social regulatory regionalism may be identified. First, in order to establish a regional regulatory integration approach, Winters (1994) argued – in the context of the EU – that the block must adopt the most precautionary position of the most hesitant, anti-integrationist member. This implication is especially important when considering integrating the various regulations for the risks of new technologies. Secondly, as social regulations are income-elastic, regulatory regionalism is inherently exclusionist, excluding less developed countries without the same level of income (Wang and Caswell, 1997). In short, social regulatory regionalism built on social protectionism may represent a stumbling-block for multilateral regulatory integration.

In order to deal with social regulatory barriers and to develop a global or even a regional normative framework, the social perspective proposes that a coordinated deeper approach to regulatory integration must be employed. The regulatory coordination strategy of international integration is associated with the 'fair-traders' school of trade theory (van Scherpenberg and Thiel, 1998). In contrast to the regulatory-competition strategy of the 'free-traders', this strategy supports bilateral, plurilateral or multilateral efforts to level the social regulatory playing-field. This is predicated on the belief that centralized regulatory approaches enhance welfare by explicitly addressing social dimensions, rather than leaving them to the forces of economic competition. Crucially, regulatory coordination is a conciliatory or cooperative approach to overcoming divergent social regulatory barriers, emphasizing shared objectives and establishing social regulatory floors – preventing a social regulatory race to the bottom. If deeper social integration cannot be pursued, then the social perspective supports a rejection of international integration based only on regulatory competition. For instance, the proposed Multilateral Agreement on Investment in 1998 and the launch of a round of trade negotiations at the WTO Ministerial Meeting 1999 in Seattle were prevented in part by the influence of social interest groups who refused to support an integration agreement that was limited to international economic integration only.

In short, the social perspective suggests that regulatory development should focus on technological precaution and social responsiveness according to a social-rationality approach, while regulatory integration should pursue deeper social integration according to the regulatory-coordination paradigm.

1.2.3 Assessing regulatory development and integration

An important similarity between the two perspectives discussed above is that the nation-state retains social regulatory autonomy or sovereignty. Yet Table 1.2 reveals crucial differences between the economic and the social perspectives. The economic perspective supports regulatory development and integration focused on economic principles of market efficiency and technological progress through a depoliticized and decentralized process of international economic integration facilitated by regulatory competition. The social perspective supports regulatory development focused on the non-market public preferences, expectations, concerns and fears and regulatory integration facilitated by a deeper regulatory coordination strategy.

The challenge of social regulatory barriers to traditional trade diplomacy emerges because the economic perspective has dominated trade diplomacy, while the social perspective dominates social regulations, such as food-safety and environmental-protection regulations. Given the discussion above, it is clear that assessing the trade challenge of social regulatory barriers and regulatory regionalism first requires an understanding of how regulations have developed within a jurisdiction, because the domestic regulatory approach sets the prospects of and limits to regulatory integration. The objective now is to examine ways to encourage the development of a stable regulatory framework and to identify an effective regulatory integration strategy that appropriately acknowledges the social dimensions embedded in regulatory barriers.

In respect of developing a stable regulatory framework and specific regulations within this framework, a jurisdiction must find some way to balance the scientific-rationality approach of the economic perspective and its calls for technological progress, with the social-rationality approach of the social perspective and its calls for technological precaution. Can the RAF be used to achieve this balance? Despite the polarity between the two perspectives, it will be argued that the scientific-rationality and the social-rationality approaches may be integrated to form an effective risk-analysis approach. For instance, Beck (1992) argued that 'scientific rationality without social rationality remains empty, but social rationality without scientific rationality remains blind'. Building on this argument, this study will explore ways of employing the RAF in the development of GM-crop regulations in a manner that effectively acknowledges the competing interests and balances technological progress with technological precaution.

In respect of creating an effective integration strategy simultaneously accounting for both the economic and the social perspectives, a regulatory jurisdiction faces three integration parameters (Fig. 1.1). First, the regulatory jurisdiction must decide the level of integration to be pursued: regional (bilateral or plurilateral) or global (multilateral). Secondly, the regulatory jurisdiction must choose the depth of integration to be pursued: shallow economic integration or deeper social integration. Thirdly, the regulatory jurisdiction must choose what type of integration strategy to pursue: the regulatory-competition strategy or the regulatory-coordination strategy. The various integration approaches, along with some examples of each, are summarized in Table 1.3; however, it is illustrative to note that the traditional trade-diplomacy approach of the current international trade regime would be located in IA in Fig. 1.1 .

There are two crucial factors linking regulatory development within a jurisdiction to the regulatory integration strategy pursued by a jurisdiction. First, the traditional regulatory role of the state will

Table 1.2. Comparison of economic and social perspectives.

	Economic perspective	Social perspective
Regulatory development	Scientific-rationality approach Focus: correcting market failure Technological progress	Social-rationality approach Focus: social responsiveness Technological precaution
Regulatory integration	Economic integration (shallow) Strategy: regulatory competition	Social integration (deeper) Strategy: regulatory coordination

1. Integration Level

I. Global/multilateral	II. Regional/bilateral or plurilateral

3. Integration strategy

2. Integration depth		Regulatory competition	Regulatory coordination
	Shallow 'economic' integration	A	B
	Deeper 'social' integration	C	D

Fig. 1.1. Integration parameters of the regulatory jurisdiction.

influence the regulatory framework developed (Woolcock, 1998) and this influences the integration approach pursued. Generally, if the regulatory tradition is closely associated with the economic perspective, then it is likely that shallow economic integration based on regulatory competition will be the preferred integration approach (area A in Fig. 1.1). In contrast, if the regulatory tradition is closely associated with the social perspective, then it is likely that deeper social integration based on regulatory coordination will be the preferred approach (area D in Fig. 1.1).

Secondly, the competitiveness of the jurisdiction is important in determining the regulatory development and integration strategies. If, for instance, the regulatory jurisdiction enjoys a commercialization lead and competitive advantage in a particular regulatory area, then it is likely that it will have well-developed regulations that domestic firms have already internalized, so that the regulations of other jurisdictions will not be difficult to comply with. In this case, it is likely that the regulatory jurisdiction will support international economic integration according to regulatory competition. Alternatively, if the regulatory jurisdiction is at a commercial lag or competitive disadvantage *vis-à-vis* other jurisdictions, it may be unlikely that it will have regulations developed enough to deal with the influx of advanced foreign technologies. In this case, it is likely that the regulatory jurisdiction will support social integration according to regulatory coordination, in order to be protected from the new technology and the competitive foreign products.

Clearly, there is a synergistic relationship between these two factors. A regulatory jurisdiction may enjoy a commercialization lead, because its traditional regulatory role supported technological progress and this technological progress may have brought tangible benefits that encouraged a regulatory framework even more congruent

Table 1.3. Summary of integration parameters: approaches and examples.

A	Aim is to facilitate economic integration according to the national treatment principle in order to allow decentralized market activity to achieve economically efficient and optimal outcomes. This is the approach employed in international and regional economic analysis of customs unions and free-trade areas (Grimwade, 1996; Jovanovic, 1998a–e). I Multilateral: the traditional trade-diplomacy framework exemplified by the GATT/WTO trade regime II Regional examples include: Association of South-east Asian Nations Free Trade Area (ASEAN FTA); North American Free Trade Area (NAFTA); Caribbean Economic Community (CARICOM); the former European Free Trade Area (EFTA); Mercad Comun del Sur (MERCOSUR); and the 1995 proposal for a Transatlantic Free Trade Agreement (TAFTA)
B	Aim is to facilitate economic integration (multilateral or regional) by coordinating divergent regulations in order to develop a common framework, thus removing market failures and access barriers and allowing decentralized markets to achieve economically efficient and optimal outcomes. I Multilateral examples include the WTO Agreement on Sanitary and Phytosanitary Measures, which attempts to remove domestic regulatory barriers for food safety by coordinating regulations within international standards-setting organizations, such as the Codex Alimentarius (see Chapter 3) II Regional examples include: the EU–US Transatlantic Business Dialogue (TABD), 1995, which is a forum for commercial interests to identify regulatory barriers; the Canada–EC Framework Agreement for Commercial and Economic Cooperation, 1976; and the EC–Canada Trade Initiative (ECTI), 1998
C	Aim is to transform the market to 'internalize' or appropriately value traditionally non-market, social objectives, so that decentralized market activity can achieve efficient and optimal outcomes. I Multilateral examples include: academic literature on food economics (Caswell and Henson, 1997; Caswell, 1999; Perdikis *et al.*, 1999; Spriggs and Isaac, 2001), environmental economics (Siebert, 1991; Paul, 1996; Swanson, 1997) and labour standards (Langille, 1996) II Regional examples include: academic literature on corporate law (Carney, 1995; Bratton and McCahery, 1996; Romano, 1996) and the New Transatlantic Marketplace (NTM) initiative, 1995, and the follow-up Transatlantic Economic Partnership (TEP) initiative, 1998, which both outlined a transatlantic commitment to liberal market objectives as a stepping-stone for multilateral integration
D	Aim is to develop a global or regional normative social construct through cooperative, *ex ante* coordination of domestic economic and social regulations. I Multilateral examples include: the overall mandate of the United Nations (UN); the UN Environment Programme (UNEP)'s Biosafety Protocol (Isaac and Phillips, 1999b,c; see also Chapter 4, section 4.2); multilateral environmental agreements such as the Kyoto Protocol and the Basle Convention on Hazardous Waste (see Chapter 3, section 3.2.2.A); and efforts of international organizations, including the International Labour Organization (ILO) II Regional examples include: the EU Eco-Labelling Scheme (Isaac and Woolcock, 1999a); the Transatlantic Consumers Dialogue (TACD), 1998; the Transatlantic Environmental Dialogue (TAED), 1998; the proposed 1995 EU–US Transatlantic Treaty; the Transatlantic Declaration on EC–Canada Relations, 1990; and the EU–Canada Joint Political Declaration and Action Plan, 1996

with technological progress. Moreover, as a result of these two factors, there is a path-dependency aspect to the development of a regulatory framework and the subsequent support for a regulatory integration strategy. The various political-economy factors (i.e. the economic and social interests) within a regulatory jurisdiction establish a trajectory for the regulatory framework that determines the type of regulatory integration strategy that can be adopted. Deviating from this trajectory can be very difficult.

The integration approaches and examples summarized in Table 1.3 are generalizations, because, in reality, it is difficult to find a regulatory integration example that fits entirely into each category. Perhaps the most illustrative example is the regional integration of the European Union (discussed in greater detail in Chapter 7). The EU-style regulatory integration, set out in the 1985 White Paper on the Single European Market (European Commission, 1985), simultaneously pursues both economic and social integration according to a blended strategy of regulatory coordination and regulatory competition (Woolcock, 1996). In fact, the EU-style integration straddles IIC and IID in Fig. 1.1.

Yet, despite its generalized nature, this conceptual model of integration is useful in capturing the complexity of the task of addressing social regulatory barriers. It reveals that traditional international trade diplomacy (IA in Fig. 1.1) is actually only a very narrow approach to regulatory integration, which implicitly supports an economic perspective on regulatory development. Moreover, it deliberately omits a significant array of issues and concerns raised by social regulatory barriers; in particular, it omits the social perspective. As social regulatory barriers are inextricably linked to the social perspective and to the regulatory development process, this omission is unsustainable.

1.3 Social Regulatory Barriers: a Case-study

This study examines this fundamental shortcoming of traditional trade diplomacy through a case-study of a current and complex issue – the transatlantic trade of GM crops. At the core of the transatlantic differences is a conflict between the economic and the social perspectives on regulatory development and integration. The thesis is that social regulatory barriers can only be addressed through a regulatory integration strategy that acceptably accounts for both perspectives. In other words, in order to remain a viable force in international integration, trade diplomacy must be amended. Yet this study also goes beyond just comprehensively identifying a problem. In Chapter 8, it also brings together the previous sections in order to propose a trade-diplomacy approach that, although reaching deep into areas of national competence, is congruent with a stable, operational and socially acceptable regulatory framework, capable of overcoming regulatory regionalism and social regulatory barriers to trade.

2 Agricultural Biotechnology

Modern biotechnology represents very sophisticated technological innovations at the frontiers of science, embedded with economic and social implications. It involves techniques capable of altering the functions and characteristics of living organisms. As its application across medical, pharmaceutical, chemical, forestry, fishery, environmental and agricultural uses is rapidly growing, the potential economic implications of modern biotechnology for the industrial landscape are enormous. Specifically, modern agricultural biotechnology has already been used to alter the function and characteristics of agricultural crops and considerable research is under way that is aimed at applying the techniques to more varieties of agricultural crops. Yet, while applications in other sectors have been readily accepted, agricultural biotechnology has been controversial.

The objective of this chapter is to define what is meant by genetically modified (GM) agricultural crops, to describe both current and future applications and to assess the factors that have made the consumer acceptance of GM crops controversial. There are two important caveats. First, this description is not intended as a comprehensive introduction to biotechnology (see Krimsky and Wrubel, 1996; Grace, 1997; Ho, 1998; Wartburg and Liew, 1999; McHughen, 2000). Secondly, the science is discussed to the extent necessary to provide background to the regulatory policy debates associated with the development and integration of regulations.

2.1 Modern Biotechnology and Agricultural Crops

Although this study is focused on the social implications of science, it is necessary to provide a description of the science of modern agricultural biotechnology because it is clear that there is a broad ignorance of what GM crops are and, perhaps more importantly, what they are not. While every effort is made to keep this section brief, an understanding of GM crops is a crucial prerequisite for any credible assessment of appropriate regulatory approaches.

2.1.1 The science

Enhancing the desirable characteristics of agricultural crops and limiting the expression of undesirable characteristics have always been objectives of agricultural production (Kenney, 1986). These attempts developed into sophisticated techniques of plant breeding, with the result that highly specialized crops were developed to meet a diverse range of human needs and ecological conditions. Although successful in developing increasingly useful agricultural crops, plant-breeding techniques remained time-intensive and plant breeders lacked the specific knowledge of how characteristics were really expressed in the crops. As a result, there was a significant degree of uncertainty in the trial-and-error approach to varietal development. For instance, the hybridization of maize through traditional plant breeding took 20 years and involved several iterative steps (Harlander, 1993). First, two parental varieties with desirable characteristics were crossed to produce seeds. The offspring were then grown out in field trials to identify those plants that expressed the desired characteristic, while those that did not were discarded. Next, the desirable offspring were backcrossed with the parents to further isolate the desirable traits. This trial-and-error approach to traditional plant breeding continued until a desired hybrid variety was produced. In this process, there was a lack of precision and sterility was common in the hybrids.

Modern agricultural biotechnology is differentiated from traditional plant-breeding techniques, because it combines the knowledge of the role of genetics in the expression of characteristics with the techniques and procedures capable of modifying the genetic make-up in order to modify the characteristics. This is primarily the role of molecular and cellular biology, where the former is 'the study of DNA' and the latter is 'the study of the structure and function of living cells' (Barton, 1998).

The key to genetic modifications is that all organisms interpret DNA in the same way. In this sense, all organisms are related (Office of Technology Assessment, 1984; Barton, 1998). However, sexual compatibility for the most part limits genetic transfer. Traditional plant-breeding techniques, such as the hybridization of maize, attempted to isolate the expression of desirable characteristics by controlling the sexual reproduction of crops. Modern biotechnology allows plant breeders to isolate and control genetic traits at a much more specific level. While the term genetic modification has been used widely to refer to all biotechnologies, there is, in fact, an important distinction. In genetic modification, the DNA of an organism is altered so as to produce some desired result but no 'foreign' DNA is added. In transgenic modification, DNA is actually transferred between organisms. Sequences of DNA are isolated in an organism, using techniques of molecular markers, and are cut from an organism using restriction enzymes. These enzymes recognize certain sequences of DNA according to their nitrogenous bases. Once the pieces of DNA are cut from an organism, they must then be pasted into the DNA of another organism. There are three techniques for importing and pasting specific pieces of DNA (Fincham and Ravetz, 1991). The first technique employs a bacterial vector, such as the common *Agrobacterium* technique, where desirable DNA is pasted into the bacterium, which then transfers it to the inside of the target cell. Similar in operation is the second technique, employing a virus vector. The disease symptoms of the virus are removed, but not its function for initial infection and replication. The third technique is vectorless, such as the 'gene gun' where tungsten bullets, coated with the desirable DNA, are shot into the target cell. In contrast to the sexual transfer of the entire genetic blueprint all at once, as in traditional plant breeding, transgenic modifications can be quite precise and can circumvent the sexual-compatibility limitation (Cape, 1986). Despite the distinction between genetic and transgenic modifications, this study follows the standard practice whereby genetic modification will be used synonymously with transgenic modification unless otherwise specified.

When the genetic modification is complete, other biotechnology techniques may be employed to assist the development of the genetically modified organism (GMO), such as tissue-culture techniques and gene-mapping or gene-tracking techniques (Fincham and Ravetz, 1991; Harlander, 1993). Tissue-culture techniques are essentially a conventional practice of traditional plant breeding, but with some improvement. The cells from the genetic modification are cultured into seeds, and contained growth trials are performed to assess the viability of the GM variety. This is followed by controlled field trials performed to assess whether or not the GM variety expresses the desired characteristics. Gene-mapping and gene-tracking techniques allow plant breeders to identify if the desired genetic modification is present in the target cells and seeds without having to grow out the seeds in field trials (Barton, 1998). One common, but controversial, method is to use an antibiotic-resistant marker gene to identify the desired genes. Cells are cultured in an environment of antibiotic, and only those that survive would have retained the gene.[1] Gene mapping both decreases the need for field trials and decreases the risk of release of unwanted GM varieties, because the desired traits will be identified in the seeds prior to the field trial.

The point is that modern biotechnology is really a package of techniques allowing for the varietal development of agricultural crops at a more precise and perhaps more controlled level than ever before. It is a process or production technique with the power to alter conventional genetic processes, creating GM varieties of traditional agricultural crops or creating novel GM varieties never before characterized.

This is not to say that there is no uncertainty with these techniques. Of course, there still exists considerable uncertainty with gene functions, but this uncertainty also exists with traditional plant breeding. Modern biotechnology, however, offers the plant developer far more control over varietal development than that possible with traditional plant-breeding techniques because the genes are known to the developer and can be tracked in the cells of the GM crop. In fact, due to this precision, it has been argued that genetic modifications are safer and more controlled than traditional plant-breeding techniques, which manipulate the entire genome of the parents, rather than just specific genes (Harlander, 1993; Hullar, 1993; van den Daele *et al.*, 1997).

2.1.2 Current and future applications of agricultural biotechnology

The application of modern biotechnology to agricultural crops can generally be categorized into three types: production-trait applications, output-trait applications and applications to create bioengineered products (Brenner, 1998).[2] These three categories will be described below. The first type of application is currently the most widespread, while output-trait applications or bioengineered products are indicative of the future of GM crops.

Production-trait applications of agricultural biotechnology represent a scientific response to long-standing agricultural problems, which had traditionally been addressed through domestic agricultural-

[1]Fincham and Ravetz (1991) argue that, although the antibiotic-resistant marker gene has been the most widely used, it is not the only gene-mapping technique available. Others include the polymerase chain reaction (PCR) technique and β-glucuronidase (a blue pigment for 'visual' mapping).

[2]Other categorizations include: Zilbermann *et al.*'s (1997) supply-enhancing, pest-control, quality-modifying, new products; and Shimoda's (1997) input traits, output traits, performance traits. Although different, these categorizations indicate that modern biotechnology can either enhance conventional traits or create wholly new ones in agricultural crops.

support programmes. Agricultural production has always been a risky venture, characterized by a significant degree of possible variation in crop quantity (yield) and quality each year. The risks to the quantity and quality of agricultural production are from the weather (e.g. drought, floods, hail and frost), from soil conditions (e.g. salinity, nitrogen depletion and erosion), from disease (e.g. rot, fungal and rust) and from pests (e.g. bacteria, viruses, nematodes, insects and animals). In both North America and Europe, various financially significant domestic support policies are employed to stabilize the agricultural sector in the face of such risks.

The attractiveness of GM-crop varieties is that they essentially provide scientific solutions to agricultural production risk through attempts to improve the production traits of agricultural crops. For instance, new GM varieties of conventional crops have been created (or are being developed) with a higher degree of stress tolerance to ecological conditions and with a higher degree of resistance to pests and disease.

Two of the most common production-trait modifications are herbicide tolerance and insect resistance, traits that have been developed for use within the intensive agricultural system. In respect of herbicide tolerance, crops have been genetically modified with a gene found in a soil bacterium that is able to metabolize (digest) the non-selective, broad-spectrum herbicide gluphosinate, rather than be destroyed by it (Moschini *et al.*, 1999). In respect of insect resistance, several agricultural crops, such as maize and cotton, have been genetically modified to express the pesticidal characteristics of *Bacillus thuringiensis* (Bt), a soil microorganism that produces a protein toxic to certain insects (Harlander, 1993).

Production-trait applications were the most common GM crops up to the 2000 crop season, with over 100 million acres planted to GM varieties (James, 2000). Of these applications, most are single-trait stacking modifications, whereby the genetic material for, say herbicide tolerance, is transferred, creating a GM variety that is herbicide-tolerant. The two most frequent single-trait stacking modifications, were for herbicide tolerance and insect resistance. However, multitrait stacking modifications represent the future of production-trait GM varieties (Brenner, 1998) – that is, transferring the genetic material for, say herbicide tolerance, insect resistance and virus resistance, to one plant organism, creating a GM variety that simultaneously expresses the three desired traits. At the same time, new agricultural crops will be subject to production-trait applications. Therefore, production-trait applications will both deepen, with multitrait stacking, and widen, to include new crops without current GM varieties.

There are two important aspects of production-trait applications. The first is that they do not require the adoption of new agronomic practices or farm implements. They can be produced within the traditional intensive agricultural production system, although they may require changes in the chemical regimes. The second is that, since the end attributes of the GM varieties remain the same or 'substantially equivalent' to those of conventional non-GM varieties, they are both for the same end use and sold into the same processing and distribution system, where it is virtually impossible to distinguish between the two.

The second broad type of GM crops are output-trait applications, targeting those commercial characteristics of the crop that determine its value, in order to increase value by increasing the expression of the desirable characteristics. Improving crops to enhance value may be viewed as a scientific response to the problem of crop quality. GM varieties of crops are being customized to meet the specific demands of end-users, such as livestock feeders, food processors or industrial users, who may place a premium on high-quality products. High-quality characteristics include improved nutritional content (e.g. protein and oils) or flavour or the functionality of the crops, such as delayed ripening or rotting. Functionality can also include enhanced processing characteristics, such as ease of separation of fibres, oils, starches, sugars and proteins (Brenner, 1998), where increased processing ease would translate into decreased energy requirements (Bonny, 1999).

Similarly to the production-trait applications, output-trait applications do not require the adoption of new agronomic practices or massive investment in new agricultural implements. These varieties may be produced according to traditional agronomic practices. Unlike production-trait applications, however, they do require changes in the distribution of agricultural commodities. These varieties have end attributes that need to be differentiated from those of the conventional varieties in order to capture the value premium. Output-trait applications create incentives for more active management of the crop distribution system through segregation. However, this degree of specificity is not yet part of the supply chain. In fact, in North America, improved output traits made up less than 1% of total acreage of GM crops in 1997, 1998 and 1999 (James, 2000), in part because the bulk nature of the agricultural commodity distribution system makes it difficult to ensure segregation between the desired varieties and other varieties without some sort of price premium.

The third broad type of agricultural biotechnology application is to create bioengineered products. Brenner (1998) suggests that, with bioengineered products, 'the power of sunlight and plant physiology are harnessed to replace expensive chemical synthesis processes'. Such applications would have industrial uses far beyond those of traditional agricultural products. Yet Brenner (1998) also notes that such applications demand a much more advanced level of biotechnological sophistication than that of current generations of single- and multitrait stacked GM varieties.

Bioengineered GM varieties would be entirely novel, rather than just improvements to conventional varieties. For instance, in the pharmaceutical sector, such applications, known as 'pharming', would allow agricultural crops to be used as biofactories producing high-value pharmaceuticals or edible vaccines that are currently produced using relatively expensive chemical synthesis processes. Such applications would have the further benefit of a decreased use of synthetic chemicals. Also, agricultural biotechnology may be employed for high-tech nutritive fortification of foods designed for health care and disease prevention, essentially becoming nutriceuticals. For instance, a current research initiative involves the transgenic modification of potatoes with cholera B toxin. Consumption of the novel GM product creates the production of human cholera-resistance antibodies. Other edible vaccines include GM crops with enhanced cancer-fighting antioxidants, probiotics and prebiotics. Another example is the nutritionally enhanced vitamin A GM rice developed to address the serious vitamin A deficiency in Thailand and other South-east Asian countries. Finally, an entirely industrial application of agricultural biotechnology would be the creation of plant-based polymers replacing petroleum-based polymers currently used in synthetic fibres, plastics and even fuels (Brenner, 1998). An example of this is a plant-based credit card developed by Monsanto and initially endorsed by Greenpeace Europe as an alternative to plastic credit cards (*Globe and Mail*, 1998). The potential benefit is the creation of completely biodegradable polymers.[3]

Unlike production- or output-trait GM varieties, novel bioengineered products would require substantially different agronomic practices. Also, to ensure that GM crops grown for non-food uses are kept out of the food supply, bioengineered products would require an effective segregation and identity-preservation production system.

Despite the potential of output-trait applications and bioengineered GM products, there are current limits to what agricultural biotechnology can achieve. For instance, it has been argued that 'no scientists will be able to take a tomato, add 20 genes from a cow and 30 genes from wheat, and come up with a crop that has the nutritional qualities of beef and bread' (Brill, 1988). Further, the challenge facing crop developers is not how to prevent GM crops from getting out of control, but, rather, how to develop GM crops that are viable enough to grow within the agricultural system – the very same problem that faced traditional plant breeders.

[3]Even one of the most ardent critics of modern biotechnology, Jeremy Rifkin, views this potential contribution as beneficial. 'The biotech age ... holds great promise: a cornucopia of new plants and animals to feed a hungry world ... and new genetically engineered sources of energy and fibre to propel commerce and build a "renewable society"' (Rifkin, 1998a).

This brief examination of the current and future applications of agricultural biotechnology has revealed several important features. First, it appears that agricultural biotechnology is, in fact, poised to deepen and broaden its impact upon economic production. Secondly, it has been argued that agricultural biotechnology and recombinant DNA (rDNA) techniques represent another phase in the knowledge intensification of seed development. Thirdly, it has also been argued that current applications of agricultural biotechnology are not significantly novel applications. Instead, they are more modest applications, incrementally made and very much in keeping with the systematic varietal development process characteristic of traditional plant breeding.

From the point of view of regulatory policy development and integration there are four important distinctions to identify. First, not all genetic modifications are transgenic modifications. From a plant perspective, transgenic modifications, as discussed above, involve the transfer of genetic material between plants or other organisms. However, some modifications, such as antisense modifications[4] and mutagenesis,[5] only alter the genetic material within a plant's cell in order to achieve desired results, therefore there is no transfer of genetic material. It appears that the greatest opposition to GM crops is actually directed at transgenically modified (TGM) crops, where genes from sexually incompatible organisms are combined. It is important to disentangle genetic modification techniques from transgenic modification techniques, because not all biotechnology applications are associated with those concerns that are really only relevant for transgenic modifications.[6]

The second important distinction is between a GMO and a living modified organism (LMO). An LMO is a subset of a GMO in that it is a GMO that retains metabolic activity. For an example, consider GM canola/rape-seed. As a seed, it is technically both a GM seed and an LMO, because it remains capable of propagation. Crushed into canola oil, it is no longer capable of propagation and is no longer an LMO, and yet it remains a derivative of a GM crop in the strictest sense (see the discussion on the Cartagena Protocol on Biosafety in Chapter 4).

Thirdly, GM crops are not always 'novel' plants. Novel plants, known as plants with novel traits (PNTs) are those for which a natu-

[4]From Harlander (1993); antisense modifications involve 'switching off' the function of certain genes within the organism's genome in order to produce desired effects, e.g. the ripening genes in Calgene's FLAVR SAVR tomato had been selectively inactivated.
[5]From Harlander (1993); mutagenesis involves exposing seeds to a mutagenic agent (ethylmethylsulphonate (EMS)) and then growing seeds out to select those resultant plants with desired traits. In this case, the genetic modification is a more random or imprecise process than transgenic modifications, with the deliberate environmental release of less characterized plants, and yet this method has not been controversial.
[6]For an example of a failure to disentangle this important distinction, see Sheppard (1997), where the concerns really associated only with transgenic modifications are cast across all GM techniques.

rally occurring counterpart does not exist. PNTs may be created either through the use of biotechnology or through traditional plant-breeding techniques. Hence, novel does not imply the use of biotechnology. GM crops do, however, imply the use of biotechnology, but not every genetic modification creates a PNT. For instance, if genetic modification is used to develop a new maize variety from two parental varieties, then the resultant GM maize is not novel, in the sense that it does not express traits never before characterized in maize varieties. Instead, it has simply enhanced maize traits that have been combined from the parents using genetic modification techniques.

Fourthly, and arising from the third distinction, GM crops do not always produce GM foods. For example, the oil and lecithin of oilseeds, such as soybean and canola, are used widely in food processing; however, oil and lecithin do not contain DNA or protein. So, although they may be derived from GM varieties, they do not contain GM material and subsequent foods produced with these inputs are not GM foods. This distinction is made clear by an examination of the difficulties of testing a food ingredient or product for GM material (Hanley and Johnson, 1999). From a technical testing perspective, a researcher can test either for the GM DNA sequence or for the presence of the introduced protein encoded by the GM DNA. In the former, the test is accurate but sophisticated and time-consuming, as the investigator must know exactly what GM DNA sequence to look for. In the latter, the test is rapid but less sophisticated and less accurate, as it relies upon the binding of an antibody to the introduced protein. The problem is that food processing easily breaks down the protein and can also degrade the DNA to the point where it can no longer be identified as a GM food. In fact, foods that have been processed (e.g. heated, fermented, acidified, extruded or highly refined) generally have no GM DNA left in them or at least the GM DNA is highly degraded. If the GM DNA is no longer in its unique sequence encoding for the particular protein, then there is no risk that a consumer is ingesting a harmful protein resulting from the genetic modification. An exception, of course, is crops that are eaten raw or unprocessed – for example, GM maize or GM tomatoes. In these two examples, the GM crops would produce GM foods. The point is that, with the exception of foods eaten raw or unprocessed, the general term ‘GM foods’ is often applied inaccurately and inappropriately.

The four distinctions are important in considering how to regulate GM crops appropriately. For instance, if public concern really lies with the transfer of genes between sexually incompatible organisms – transgenic modification – then regulations should target TGM crops, not all GM crops. If it is plant novelty that is the concern, then novel-based regulations are more appropriate than technology-based regulations. If the concern lies with the protein structure of GM organisms, than the focus should be only on those products that still contain the GM DNA sequences. The distinction between GM crops and LMOs is crucial, because the Cartagena Protocol on Biosafety is an international treaty governing the transboundary movement of LMOs. Accordingly, unless GM crops are shipped in seed form and capable of propagation, they should not fall under the regulatory principles of this protocol. Therefore, understanding these important distinctions is vital in establishing regulatory approaches that respond to actual consumer concerns rather than approaches built on vague, ambiguous fears about a misunderstood application of modern agricultural biotechnology techniques.

2.2 Agricultural Biotechnology: an Overview of Consumer Acceptance

Up to this point, the application of modern biotechnology to agricultural crops has been examined as a scientific, technological innovation that promises to have vast implications for agricultural production. Yet, at the most basic conceptual level, all production is for consumption and some consumers have not accepted GM crops, because they remain unconvinced about the consumer benefits or because they have concerns or fears about the technology.

In understanding the rejection, it is useful to remember that consumers are not just economic agents but also social agents who vote and participate as citizens, ultimately shaping the national social and political context within which economic forces operate. For instance, venture capitalists may not be willing to fund GM-crop developers if it would be publicly unpopular.[7] Similarly, producers may be unwilling to plant GM crops either if they cannot market them or if they will be vandalized by eco-warriors. More importantly, however, consumers have enormous influence over the research, development and commercialization of GM crops through their influence upon regulatory approaches. Therefore, a critical assessment of the social implications of agricultural biotechnology must focus on the issue of consumer acceptance and its role in shaping the domestic regulatory approach and the subsequent regulatory integration strategy.

The objective of this section is to identify the predominant issues associated with the consumer acceptance of GM crops, because these issues are cited by various interest groups active in the regulatory

[7]In 1999, Deutsche Bank Europe predicted that biotechnology firms will become a 'pariah' of shareholders and will suffer an 'earnings nightmare', while GM crops will be a liability for farmers. Also, the Public Ledger (1999a) reports that Monsanto's shares slumped from a high of US$51 in May to US$38 in October 1999, because of shareholder concerns about consumer attitudes towards GM crops.

development process. While the issues are presented here in a descriptive fashion, they will be the basis for assessing the economic (Chapter 3) and the social (Chapter 4) interests influencing GM-crop regulations.

2.2.1 Agricultural biotechnology and consumer concerns

There are four types of consumer concerns: economic concerns, human health and safety concerns, biodiversity concerns (including animal, plant and environment health) and moral, ethical and/or religious concerns. A synergistic relationship exists between the four types of consumer concerns. They may be positively related and mutually reinforce a particular consumer position. Inversely, they may be negatively related, forcing the consumer to strike a balance between benefits, on the one hand (e.g. lower price or improved nutritive content), and costs, on the other (e.g. a moral perception that genetic modification is unnatural or wrong). The four types of consumer concerns are discussed below.

A. Economic concerns

Economic consumer theory presents consumers as economic agents, driven by the principle of non-satiation, who consume normal goods based upon the attributes of the goods and subject to a budget constraint. The attributes of the good – price and quality – are relative variables that assist the consumer in choosing the consumption bundle that maximizes consumer welfare or utility. Much economic analysis is focused on the price to determine the cost/benefit to consumers from the use of GM crops. If the use of a GM crop reduces the relative price of a normal good, then, according to economic consumer theory, the consumer will choose to consume more of that good and, subsequently, consumer welfare or utility is increased through the use of GM crops (Hoban, 1996c; Moschini *et al.*, 1999). Similarly, if the use of GM crops leads to increases in the price of that good then consumer welfare or utility is decreased (Giannakas and Fulton, 2000).

Economic consumer concerns may also involve concerns about the broader economic impact of GM crops, positively or negatively influencing acceptance. For instance, a high concentration of research capacity, providing well-paying high-technology jobs and creating economic spillover, may enhance the perception that GM crops are associated with wider economic benefits. Similarly, improvements in productivity and income among the rural sector may also be considered by consumers to be an economic benefit, positively influencing consumer acceptance. Inversely, the broader economic impact of GM crops may negatively influence consumer acceptance. Three examples are illustrative of this. First, consumers may view GM crops as facilitating the further industrialization of agriculture and being socially destructive because of the economic displacement of rural communities.[8] Secondly, consumers may perceive that all the economic benefits are accruing to the large, private, multinational firms commercializing GM crops or to farmers in the intensive agricultural sector, with no benefit to consumers from the new technology. Thirdly, consumer concerns may be associated with the potential economic impact of GM crops used for import substitution upon developing countries.

B. Human safety and health concerns

Apart from economic concerns, consumers are concerned about the human safety and health implications of GM crops. Human safety concerns refer to the short-term absence of illness immediately after consumption, while human health concerns refer to the longer-term health implications, such as cumulative nutritional impacts. Essentially, the fear is that genetic modifications will result in toxigenic, pathogenic, infective or invasive changes to the plant, affecting human safety and health. Also, a secondary human health concern associated with GM crops arises from the use of antibiotic-resistance genes in gene-tracking procedures. It is a secondary concern because, although there are no direct consequences for human safety or health from consuming such marker genes, the concern is that their use will increase the incidence of antibiotic resistance in bacteria that are harmful to humans. Such concerns have a negative impact on consumer acceptance.

Yet GM crops with improved nutritional characteristics, such as the cholera B potato or the vitamin A rice, can have positive impacts upon public health and may enhance consumer perception of agricultural biotechnology. Further, GM crops eliminating or reducing the need for herbicides and pesticides would have significant impacts upon human safety, health and various allergies and sensitivities.

C. Biodiversity concerns

There are also consumer concerns associated with the potential impact of GM crops upon biodiversity, which can positively or negatively influence consumer acceptance. From a biodiversity point of view, a current international focus is on sustainable development,[9]

[8]Cited by Britain's Prince Charles as a negative aspect of agricultural biotechnology (*The Times*, 29 October 1998).

[9]The Brundtland Report defines sustainable development as 'development that meets the needs of the present without compromising the ability of future generations to meet their own needs' (WCED, 1987).

which for agricultural purposes translates into sustainable agricultural production techniques. While GM crops can have negative impacts on sustainable agriculture and biodiversity, they can also have positive impacts (Bonny, 1999).

Biodiversity concerns involve the perception that GM crops will result in aggressive crops or crop relatives that disrupt the ecosystem. For instance, one argument suggests that the transgenic modifications could transfer to either conventional non-GM varieties of the same plant (the so-called invasion-of-origin concern) or to non-target plants and organisms through vector-mediated horizontal gene transfer and recombination. It has been argued that genomes are dynamic and predisposed to incorporate foreign DNA (Wheale and McNally, 1990). In fact, some observers have made dire predictions of genetic pollution, habitat destruction and the destabilization of entire ecosystems brought on by aggressive GM crops (Ho, 1998; Rifkin, 1998a). Other biodiversity concerns are associated with the fear that farmers producing herbicide-resistant GM crops will apply herbicides in a reckless, irresponsible fashion in an attempt to control weeds, harming biodiversity (Consumers' Choice Council, 1999).[10] Such concerns negatively influence consumer acceptance of GM crops.

In respect of the potential biodiversity benefits of GM crops, it has been argued by many supporters that the future of GM crops is essentially chemical-free production congruent with sustainable-farming trends. Agricultural production free of herbicides, pesticides and fertilizers would have enormous benefits on biodiversity. GM crops, tailored to meet the ecological conditions of various regions, could increase the range of alternative crops and crop varieties that producers could choose to plant. This development would break the trend of monocropping and result in increased biodiversity. As already discussed, the techniques of gene mapping and gene tracking can provide greater control over crop development and effectively limit the number of field trials required to determine if the phenotypic characteristics are present or not. Further, GM-crop developers are exploring ways to lock in or turn off the transferred genes in GM crops to prevent genetic drift. Such developments should positively influence consumer acceptance of GM crops.

[10]See also the comments of Britain's Prince Charles (*The Times*, 29 October 1998), which seem to reveal a contradiction. On one hand, he argues that the farmer is the responsible steward of the land, while on the other, he argues that farmers cannot be trusted with herbicide-tolerant crops because they will apply herbicide in a reckless fashion. However, in general, this particular concern about the abuse of herbicides does not seem very robust for two reasons. First, herbicide use is a significant input cost for producers, who will always seek to minimize this cost instead of overspraying in a reckless fashion. Secondly, in most cases, the wealth of the producer is embedded in the agricultural land and it is unlikely that the producer will recklessly jeopardize that wealth.

D. Moral, ethical and religious concerns

For some consumers, the acceptance of GM crops is also associated with moral, ethical and religious concerns, because these crops essentially represent the modification and manipulation of the processes of life. Survey evidence both from North America (Canada) (Einseidel, 1997) and from Europe (Eurobaromètre, 1997) suggests that consumer acceptance of modern biotechnology broadly and GM crops specifically is significantly influenced by perceptions of the moral and ethical aspects of the applications. Acceptance is found to be positively associated with those applications that are perceived to be morally beneficial. The Eurobaromètre (1997) study concluded that:

> first, usefulness is a precondition for support; second, people seem prepared to accept some risk as long as there is a perception of usefulness and no moral concern; but third, and crucially, moral doubts act as a veto irrespective of people's views on use and risk.

Further, in respect of religious concerns, the Eurobaromètre (1997) survey evidence has revealed that nearly 40% of respondents believe that religious authorities should be involved in the public policy discussions and decisions regarding biotechnology applications, while North American survey evidence has revealed that acceptance of agricultural biotechnology is negatively related, in part, to religious concern (Hoban, 1997).

The morality and ethics of the application of GM technologies to agricultural crops are also called into question because of the scientific approach pursued by the primary developers – often the large multinational life-sciences firms. In the past, scientists, through the peer-review structure of scientific investigation and disclosure, have been trusted to protect social norms in the face of new technology on behalf of the general public. Ho (1998) argues that the shift from public leadership in research on biotechnology to private leadership is associated with several substantial normative problems. First, the scientific research by private firms is 'reductionist' in nature – that is, it employs specific transgenic modifications to a variety and only assesses the impact upon that crop. It fails to assess the impact of the GM crop upon the biological system in which the crop operates – a so-called inclusionist approach. Secondly, the reductionist science supports a false dichotomy between science, as non-negotiable laws of nature, and technology, as the application of science. According to this dichotomy, science is value-free and only when it is applied as a technology do any associated normative issues emerge. This dichotomy is challenged by the argument that science is only a tool for understanding nature and that scientific research is inseparable from social norms and morals. Thirdly, the reductionist, profit-seeking motives supporting the science–technology dichotomy fail to address the public interest, since science and technology are pursued outside the public debate (often protected by proprietary claims). Only when the technology is to be commercialized is the public interest considered. Indeed, it has been argued that the profit motive means that private scientists can no longer be trusted to act in a moral or socially ethical manner (Monbiot, 1995). This argument also implies that the national economic drive to develop an internationally competitive agricultural biotechnology capacity means that government regulators cannot be trusted either. Accordingly, the result is GM crops without public consent.

Others argue, however, that having moral or ethical concerns about GM crops is a luxury enjoyed by North American and European consumers and made possible by an ample and well-distributed food supply. It has been queried whether it is more unethical to transfer genes or to allow starvation in less developed countries, where the choice is not between GM crops or non-GM crops based on moral concerns, but between living and dying (Sahai, 1997). Similarly, it has been argued that it is 'irresponsible and immoral for the well-fed to spearhead fear-based campaigns and suppress research for ideological and pseudo-scientific reasons' (Prakash, 1999). In addition, an 18-month study by a working group of the Nuffield Council (the leading UK body on bioethics) concluded that there was, in fact, a moral obligation to continue to develop GM crops because of their significant potential. It concluded that there were no grounds for a ban on GM crops, because they were not sufficiently different from conventional, non-GM crops and hence do not raise moral objections if conventional crops do not. Similarly, the Vatican's Pontifical Academy for Life stated that the use of GM technology in agricultural crops should not raise moral alarm because the advantages are greater than the risks and that the Academy does not agree with those organizations who argue that GM crops are against the will of God (*AgraFood Biotech*, 1999b).

An important feature of the four types of consumer concerns is that they tend to reflect a concern about the application, management and distribution of modern agricultural biotechnologies for specific uses, rather than a concern about the technology *per se*. This is vital in determining what type of regulatory approach is needed: a process/technology-based approach or a product/application-based approach. These issues will be discussed further in Chapter 5. The important point to make is that GM crops are inextricably linked to a broad range of public concerns. The question remains: how is consumer acceptance linked to these concerns?

2.2.2 Consumer information, trust and choice

The four types of consumer concerns identified above represent consumption parameters in the sense that consumers require these concerns to be addressed prior to making a consumption decision. The objective now is to examine the events involved in the acceptance of advanced technologies, such as GM crops.

Neoclassical economic consumer theory assumes that consumer concerns are completely addressed through the provision of information. According to the rational-choice model, the consumer has access to perfect information about all the attributes of the products including information about the inputs and the processing and production techniques, as well as the costs per unit to produce the good. Implicitly, the consumer also has access to all the information necessary to address the other consumer concerns, beyond just economic concerns.

Access to perfect information allows the consumer to make rational[11] consumption choices, creating consumer sovereignty. Consumer sovereignty is the notion that the consumer is the best judge of the ramifications of consumption in respect of her/his concerns and does not require market interventions to enhance this judgement. A consumer may be rational even without perfect information, provided the consumer chooses to be partially informed. The consumer may choose this because, for instance, the time, effort or cost required to gain complete information is not justified by the perceived benefits of being fully informed. In this case, the consumer is boundedly rational (Williamson, 1987). Generally, bounded rationality requires that the consumer trusts the partial information being provided. This suggests an important relationship between information, trust and choice. It has been argued that, essentially, 'trust can be a functional substitute for knowledge' (Eurobaromètre, 1997) in making either completely or boundedly rational consumption decisions, and consumer sovereignty depends upon the retention of choice.

Are agricultural biotechnology products congruent with the rational-choice model of consumer theory? Do consumers have access to all the information necessary to make rational consumption decisions? Do consumers trust the information providers? Do consumers always retain choice? The answer to each question is no, and hence recent negative consumer reaction to the commercialization of GM crops should come as no surprise.

Consumer goods may be categorized into three types (Tirole, 1988). First, search goods are those goods where consumers can visu-

[11]The concept of 'rational' is subject to interpretation. For instance, Rayner (1992) argues that it is a subjective term. Here it is used in the economic sense, as the efficient allocation of resources according to optimizing behaviour.

ally identify all product attributes prior to purchase and consumption. With search goods, all information about the good is effectively transferred from the producer to the consumer. Secondly, experience goods are those goods where the consumer is not able to identify all product attributes prior to purchase and consumption. There is a partial breakdown in the transfer of information from the producer to the consumer at the time of purchase. Yet the consumer can gain the necessary information through consumption experience. Thirdly, credence goods are those goods where the consumer is not able to know the full attributes of the product before or after consumption (Bureau *et al.*, 1997; Purchase, 1997). With credence goods, there is a total breakdown in the transfer of information from the producer to the consumer – an information gap. Yet what fills the information gap is consumer trust in those developing the goods (e.g. in scientists) – there is 'credence' associated with them.

Defining GM crops as credence goods is intuitive since there is a large information gap between producers and consumers (Isaac and Phillips, 1999b, 2001). This is due to two factors. First, many consumers do not understand the scientific techniques and procedures of modern biotechnology. In fact, general knowledge about genetics is often lacking among many consumers, let alone specific knowledge about transgenic modifications to agricultural crops. For instance, a recent report by the European Commission revealed that two-thirds of those surveyed did not realize that non-GM tomatoes also have genes or DNA in them.[12] Secondly, with the dominant role of the private sector, much information may be deemed proprietary and not available to consumers. Further, since the research and development that underlie the application of modern biotechnology are advancing rapidly, the information gap, created by the credence nature of GM agricultural crops, is likely to widen, not narrow.

In general, the producer–consumer information gap in not insurmountable. For instance, the credence nature of goods may be remedied through the provision of information, permitting credence goods to shift to experience and then search goods. Evidence in the USA has shown that increasing consumer information about GM crops has positively influenced consumer acceptance (James, 1997). An important element of information is transparency, as was shown in a 1998 referendum in Switzerland, where consumers/voters were asked about their position on the continued support of biotechnology research and development. Consumer support and acceptance rose as the industries applying biotechnology adopted the strategy of opening up their activ-

[12]European Commission report published in January 1999, reported in the *Financial Times* (1999a). Given such a lack of understanding of GM crops, it is remarkable, for instance, that Patrick Holden of the UK-based Soil Association describes the consumer rejection of GM crops as based on 'informed public opinion' (*Independent*, 1999r).

ities to the public. Eventually, referendum results revealed that two-thirds of 'informed' voters supported the continued research, development and application of biotechnology (European Federation of Biotechnology, 1998). However, despite the importance of transparency, it has been argued that transparency has not always been the chosen strategy of both supporters and critics of agricultural biotechnology, resulting in a dual lack of transparency (*Economist*, 1998a). Both sides have, at times, failed to be completely transparent about the opportunities and risks, exacerbating the information gap and consequently the credence nature of GM crops.

Yet simply providing transparent information will not remedy all consumer concerns. The information must be useful, but there are several challenges to providing useful consumer information. First, providing all the information necessary for consumers to completely understand agricultural biotechnology is impossible, given the scientific sophistication of agricultural biotechnology. Consumers would soon experience an overload of information (Eurobaromètre, 1997; Hoban, 1997; Chess, 1998). This occurs because there are important limitations to the consumer's ability to process information. For instance, labels cannot be expected to convey all the information that consumers may want. Indeed, it has been argued that simply labelling a product as 'genetically modified' is meaningless, because consumers want to know more contextualized information, such as which genes have been used (Grove-White *et al.*, 1997). Consumers must then rely upon the judgements of others and this gives rise to the second challenge: who should provide the useful information?

Without first-hand knowledge of the consumption ramifications of GM crops, consumers must trust that their concerns are being adequately addressed by either the industry, the regulators or a third party. In both North America (Hoban, 1997) and Europe (Eurobaromètre, 1997; Grove-White *et al.*, 1997), consumers have indicated that they most trust third parties with information about GM crops, especially environmental organizations at the international level. Trust is a delicate attribute of information providers, as it is hard to build but easy to lose (Chess, 1998). However, opinions are very diverse, ranging from those who believe in the global-welfare promise of GM crops to those who wish to see the technology completely abandoned. The polarity of opinions, producing overly sensationalized information, can leave consumers very confused.

The third challenge of providing useful information is determining who the information should target. Essentially, is it practical to try to inform every consumer, or is it more practical to target information to a few, who can disseminate it to many? It has been argued that 'instead of trying to educate the public, we should focus our attention on the media, health professionals and other opinion leaders' (Hoban,

1996c). This position recognizes that information must be accessible for the concerned consumer, and these are the information channels that they are most likely to turn to when seeking information.

A further problem remains for consumer acceptance of GM crops. Even if trusted information providers satisfactorily inform consumers, consumers may still lack consumer choice or 'informed consent'.[13] For instance, the global handling and distribution system for agricultural crops is a bulk-orientated system that involves mingling different varieties of the same crop and mingling different crops. Within this existing system, it is virtually impossible to ensure segregation of production-improved GM varieties from non-GM varieties. In order to ensure segregation, dedicated handling and storage facilities must be used, resulting in cost increases (Isaac and Phillips, 1999c). In fact, due to a purely economic decision, GM varieties approved as substantially equivalent to non-GM varieties were initially mingled in the food supply in both North America and Europe without effective segregation and labelling.

In respect of the decision not to segregate, GM-crop developers in both the North American and the European grain and oil-seed industry argued two main points. First, the GM crops had been approved as safe and substantially equivalent to non-GM varieties, so there were no safety reasons to segregate. Secondly, it was argued that the distribution system made it virtually impossible to segregate, with zero tolerance, GM from non-GM crops without significant economic costs, a view shared by both US and European industry participants.[14] Economic studies concluded that the costs of segregating would be substantial. An experimental identity-preserved production (IPP) system for GM canola/rape-seed varieties was implemented in Canada in 1995 and 1996. From this experiment, it was concluded that an IPP system created incremental costs of between C$34 and C$37 t^{-1} for grains and oil-seeds (Manitoba Pool Elevators, 1997). Other estimates concluded that developing and implementing an international IPP system would require a commodity price rise of between 140 and 180% (EuropaBio, 1997). In this sense, the decision not to segregate GM crops was not taken as a cunning strategy to push GM crops into the North American and European food supply, as some have suggested. Instead, it was made on the basis of the economic cost of developing an effective IPP system and with the view that such rises in crop prices and, consequently, food prices could not be absorbed by the industry and consumers.

[13]See also Balk (1993) for a discussion of consumer choice, referred to as 'informed consent'.

[14]See: Agrevo (1996); American Soybean Association (ASA, 1996); Central Soya Protein Group (1996); National Soybean Producers Association (1996a,b); Sparks Companies, (1996); Grain and Free Trade Association (GAFTA, 1997).

Yet, regardless of the industry's intentions to keep prices down at a competitive level, the current controversy surrounding the segregation issue is indicative of the danger of treating consumers as mere economic agents and failing to address their broader concerns.

The biotechnology industry currently faces another important decision associated with consumer information and choice, this time over labelling. Again, the stance of industry – one that is shared by Canada and the USA – is that, since labelling is not for safety reasons, economics should determine what type of labelling prevails. They argue that certifiably non-GM crops and food products should bear a voluntary label in pursuit of niche-market premiums. Consumers' organizations, however, often support labelling of any use of GM crops in the production of food products as a consumer right-to-know issue. For instance, 98% in Canada, 85% in the USA and, in aggregate, 74% of European Union (EU) consumers have all indicated at various times that they want GM labelling even if the GM crop has been approved as safe.[15] The issue of labelling will be discussed further in both Chapter 3 and Chapter 5, but the important point to make is that, despite the negative consumer reaction to the decision to ignore the broader concerns and only focus on the economic issues of segregation, it appears that some jurisdictions are willing to make the same mistake again over the labelling issue.

The discussion above demonstrates the synergistic relationship between consumer acceptance and information, trust and choice. Even if consumers are willing to accept only partial information about credence GM crops and to trust the regulators and information providers, the bulk nature of the global handling and distribution system for agricultural commodities restricts or prevents choice when segregation cannot be ensured. In this case, the consumer is unable to make even a boundedly rational consumption decision because of the absence of choice. Furthermore, this does little to ease consumer concerns. On the contrary, the credence nature of GM crops coupled with the inadequate information, lack of trust and absence of choice plays directly into consumer fears and rejection.

2.2.3 Asymmetrical consumer acceptance

Further complicating the assessment of consumer acceptance is the existence of discernible differences in consumer acceptance both across biotechnology-based products and between North American and European consumers.

[15]Based on survey summaries found in Consumers International (1999c). In particular: Canada: *Toronto Star* (1998); USA: Hoban and Kendall (1992); EU: Eurobaromètre (1997).

A. Asymmetry of consumer acceptance across biotechnology products

The asymmetry of consumer acceptance across biotechnology-based products is associated with two factors. The first is the consumer's perception of the 'primary beneficiary', while the second is the consumer's perception of control over the application. According to these two factors, modern biotechnology has been more accepted when applied to medical and pharmaceutical industries relative to agriculture.[16] Medical and pharmaceutical applications of biotechnology are perceived by consumers to be focused on human health; therefore the consumer is the primary beneficiary. When the consumer is the primary beneficiary, it has been argued that there is a greater likelihood of acceptance of the benefits and the risks (Slovic, 1987, 1990). On the other hand, GM crops with production-improved traits developed to increase yields and productivity in the intensive agricultural sector are perceived to create supply-side production benefits only. In this case, consumers are being asked to accept something that seemingly provides them with no benefits. Medical and pharmaceutical applications are also perceived to be done in controlled research facilities while the field testing and commercial release of GM crops is perceived to be uncontrolled in the environment (Hullar, 1993). Some argue, however, that the difference in acceptance of pharmaceutical or medical applications over GM crops is also related to the fact that in the former, there is a long history of stringent premarket approval processes while, in the latter, food is not traditionally preapproved in the same fashion (Horton, 1997). This implies that consumer acceptance is also related to regulatory approval. Yet Isaac and Phillips (2001) argue that an adverse circularity exists in consumer acceptance, whereby consumers will not accept products that have not been approved, but regulators – fearing a public backlash – will not approve products that are not accepted. This adverse circularity challenges the linear assumption of consumer acceptance.

Even within the broad spectrum of agricultural applications of modern biotechnology, there are asymmetries in acceptance. For instance, agricultural applications to improve human health through nutritive fortification are more readily accepted than applications to improve the commercial attributes of produce (*Economist*, 1998a). An example of a GM crop with direct consumer benefits is a GM sugar beet developed at the Centre for Plant Breeding and Research, Wageningen, The Netherlands. Researchers claim that they have developed a sugar with low caloric value because the fructans have been modified to be long-chain fructans, which are not easily digested by humans. Interestingly, this GM sugar beet was not given the label

[16]See: Hullar (1993); Einseidel (1997); Eurobaromètre (1997); Hoban (1997); *Economist* (1998a).

'Frankenstein food' in the UK media, as other GM crops were (*Metro*, 8 June 1999). In addition, North American survey results indicate that respondents are more likely to accept genetically modified fruits and vegetables than genetic modifications to livestock (Chess, 1998). This indicates that consumer acceptance is linked to perceptions of the morality or ethics of genetically modifying so-called higher-order organisms, such as animals (Einseidel, 1997; Eurobaromètre, 1997).[17]

The asymmetrical consumer acceptance across products appears to indicate an important principle: that it is the product application that matters to consumer acceptance, not the technology *per se*.

B. Asymmetry of consumer acceptance across regions

The asymmetry of consumer acceptance across regions is a particular challenge to the international trade and market access of agricultural biotechnology products, because regulatory integration efforts must acknowledge regional differences. Essentially there are no universal consumer concerns. Instead, they are shaped by historical, cultural and economic conditions (Hoban, 1997). Additionally, consumer concerns are also influenced by the current information consumers receive. For instance, in the UK, GM crops have often been inaccurately portrayed as an 'American' technology or 'Monsanto's' technology (*Ecologist*, 1998),[18] despite the fact many European firms are very active in the development of GM crops, along with European universities and public research institutes.[19] Therefore, the complicated mix of consumer concerns and asymmetries of acceptance across products must additionally be understood within a regional context. The focus of the analysis will be on asymmetries in consumer acceptance between North American and European consumers.

Broad support for modern biotechnology is greater in North America than in Europe. Hoban (1997) reports that between 66 and 75% of survey respondents in the USA indicated acceptance of biotechnology products, yet in Europe the acceptance among respondents is just over 50%. Although North American acceptance is evidently higher than in Europe, it is important to note that this acceptance was not unconditional. In fact, concern over the use of recombinant bovine somatotrophin (rbST) in dairy cows in the USA a decade ago is very similar to current European concern and action regarding the use of GM crops in the food supply. At the forefront of

[17]For a comprehensive discussion of moral and ethical problems of genetically modifying animals, see Fox (1990, 1992).
[18]See Chapter 7 for an examination of the role of the media in the UK in providing often incomplete and far from objective coverage of the issues around GM crops.
[19]For the promotion of GM crops in Europe see Chapter 7. European multinational firms have included: Agrevo (Germany), Astra-Zeneca (UK), Novartis (Switzerland), Rhone-Poulenc (France).

rbST concern was the US-based Foundation for Economic Trends (Wiegele, 1991). Due to the public concern, many industrial dairy farms refused to use rbST in their herds. Also several large food processors (e.g. Kraft USA, Borden Inc., Dannon Inc.), along with many food retailers (e.g. Kroger, Safeway, Pathmark, Stop & Shop, Vons), all boycotted milk and milk products produced from rbST herds. These boycotts remained in place until the scientific uncertainties surrounding the use of rbST had been addressed in a sufficient way to reduce consumer concerns.

To deal with the asymmetry in consumer acceptance of GM crops, both Monsanto and the European biotechnology industry association, EuropaBio, launched public information campaigns in 1998 in an attempt to increase European consumer information about the benefits of GM crops and hence increase acceptance. However, market research after these campaigns has revealed that they were very unsuccessful, as consumer acceptance in Europe decreased instead of increasing. The percentage of European consumer 'unacceptance' of biotechnology rose from 38% in October 1997 to 51% in October 1998. In Germany, the level of consumer 'unacceptance' of biotechnology was reported to be over 80% (*Financial Times*, 1999a).[20] As a result, in October 1999 the US-based firm Monsanto embarked on a consultation campaign with UK environmental and consumers' organizations, such as Greenpeace, the Soil Association, Friends of the Earth and the UK Consumers' Association (*Independent*, 1999r).[21]

Research has also indicated that, while education is positively related to consumer acceptance in North America, it is negatively related to consumer acceptance in Europe. Hallman and Metcalfe (1993) report that 80% of college-educated respondents and less than 60% of respondents with a high-school diploma or less indicated acceptance of agricultural biotechnology in New Jersey. On the other hand, research in Europe reported that consumer acceptance was negatively related to education, as acceptance was reported as lowest in Denmark, Germany and The Netherlands, which were identified as having the highest education (Almas and Nygaard, 1995).

Previous research on food-consumption trends in North America and Europe has concluded that European consumption patterns lag behind those in North America by about 10 years (Connor, 1994). The implication here is that the divergence in consumer acceptance of GM crops is just a short- to medium-term phenomenon, so that trade tensions will just disappear. However, this is not a likely conclusion in respect of GM crops, for several reasons. First, the very perception of agriculture is different in the two regions (see Part III), resulting in a

[20]See also the Consumers' Association, UK (1997) for a discussion of UK consumer acceptance.
[21]For a further discussion of UK campaigns against GM foods, see Chapter 7.

significant cultural clash. Secondly, many severe and well-publicized food-safety crises in Europe have created a cultural context of distrust in the food industry and in food regulators (Spriggs and Isaac, 2001). Food safety, in general, has become a sensitive, highly politicized issue throughout Europe, and credence GM crops, developed and commercialized by multinational corporations, appear to be just another trend to fear. In fact, it has been argued that crises and controversy create irreversible effects, implying that consumer acceptance may be on an unalterable trajectory (Joly and Lemarie, 1998). Thirdly, the politically significant environmental-protection movement in Europe has made the opposition of GM crops a main theme (see Chapters 4 and 7). Fourthly, and not to be overlooked, the commercialization lead in North America creates pressures to protect domestic biotechnology firms in European member states until they are ready to compete internationally.

2.3 Conclusions

The objective of Chapter 2 has been to define what is meant by the term 'GM crops' and to discuss why their commercialization has been so controversial. Essentially, GM crops are scientifically sophisticated, knowledge-based products. Like all new technology, they promise great opportunity, while their credence attributes raise legitimate consumer concerns about their risks. Perhaps to a degree unlike most other technologies, though, consumers have concerns beyond just economic concerns. They also have concerns about human safety and health, biodiversity and broader moral, ethical and religious concerns about the use of modern biotechnology. An information gap hinders consumer rationality, while the lack of trust and choice hinders consumer sovereignty. In this sense, it is not hard to understand why many consumers lack confidence in GM crops, demanding stringent regulatory responses.

The assessment of consumer acceptance reveals three important issues that regulatory development and integration strategies must address. First, different interests groups will focus on different concerns in an attempt to influence the regulatory development and integration strategy (as will be discussed in Chapters 3 and 4). Secondly, it appears that a common link among virtually all of the consumer concerns associated with GM crops is the focus on the application, management and distribution, rather than the technology *per se*. Thirdly, an important objective of the regulatory framework is to address a broad range of concerns about the applications of GM technology to agricultural crops, while crucially providing information, trust and choice. Regulatory development and integration strategies simultaneously accounting for these three issues will be well positioned to deal with the commercialization of GM crops in an economically and socially acceptable manner.

In Part I, a conceptual framework for analysing regulatory regionalism created by social regulatory barriers facing the trade of GM crops has been established. It is clear that, in order to understand the complexity of social regulatory barriers and the prospects and limits for regulatory integration, it is first vital to understand the regulatory development process – a complex interaction of domestic political-economy factors associated with a broad range of consumer concerns and linked to an acceptable mechanism based on information, trust and choice.

Part II
Regulatory Development and Integration

The objective of Part II is to examine the process of regulatory development in order to identify the difficulties associated with regulatory integration and the limitations of traditional trade diplomacy in respect of genetically modified (GM) crops.

In Chapter 1, it was argued that, in general, there are two perspectives on regulatory development and integration. Building on this categorization, it is proposed that the broad range of interests involved either in support or in criticism of GM crops may be categorized into economic and social interests. The former hold a predominantly economic perspective on regulatory development and integration while the latter hold a predominantly social perspective on regulatory development and integration. Of course, as argued in Chapter 1, these perspectives result in support for different regulatory development and integration frameworks. The rationales as to why the various economic and social interests support particular frameworks for the development and integration of GM crop regulations will be examined in greater detail in Chapters 3 and 4, respectively.

In Chapter 5, it will be argued that this simple categorization of interests has, in fact, a significant degree of power in explaining the contentious debates associated with the development and integration of GM crop regulations and hence the problems associated with the traditional trade-diplomacy approach. It will be argued that the debates and ensuing regulatory instability are the result of a lack of agreement between the two interests on even fundamental framework principles for regulating advanced technologies. It will also be argued that a failure to establish an international regulatory framework that could be a basis for trade diplomacy has resulted in a fragmented collection of international rules, guidelines, recommendations and codes of practice, each more or less influenced by one of the two interests. Without international leadership, two dominant regulatory frameworks have emerged: a North American and a European framework. Comparing the influence of the economic and the social interests in the development of these two dominant frameworks is the objective of Chapters 6 and 7 in Part III of this study.

3 Economic Interests

In this chapter there is an assessment of the regulatory development and integration approaches supported by economic interests. This includes the agricultural biotechnology industries, complementary agents in the agricultural sector and those farmers who have adopted genetically modified (GM) crops, as well as governmental and non-governmental organizations at various levels who tend to view the commercialization of GM crops as a positive development. Why do these interests support GM-crop technology? Generally, the economic interests argue that GM crops promise significant private and public economic opportunities and they must be viewed as a crucial component of national competitiveness. Accordingly, they support regulatory development encouraging technological progress and regulatory integration encouraging stable and predictable market-access rules for international economic integration.

3.1 Regulatory Development

In this section, the economic implications of GM crops are considered at the producer, sectoral and national level. Essentially, GM crops promise considerable private and public economic benefits, extending beyond the agricultural sector and making them an important element in the industrial competitiveness of a nation. Economic interests, in pursuit of these outcomes, support a stable and predictable regulatory framework ensuring technological progress.

At the farm level, production-trait GM crops have been developed to address important production concerns of producers. For instance, herbicide-tolerant varieties address producer concerns about weed control, by replacing many synthetic chemicals with one broad-spectrum, postemergent herbicide. Producer's costs are reduced, as they do not have to spray their fields so often. *Bacillus thuringiensis* (Bt) varieties with insecticidal characteristics reduce costs by eliminating the need for particular insecticides and the costs of applying those insecticides. Further, the economic benefits of these GM crops are enhanced by their conformity with both conventional agronomic systems and bulk commodity distribution channels, thereby eliminating the need for investment in new implements and segregation practices. Given the economic benefits, GM crops have been rapidly adopted. For instance, in the USA in 1998 20.5 million ha of transgenic crops were planted, while in 1999 28.7 million ha were planted, a growth rate in adoption of 8.2%. In fact, the global growth rate of adoption was 12.1% (James, 1999).

The next generations of GM crops, output-trait GM crops and bio-engineered products, are intended to secure farm-level economic benefits by exhibiting qualities demanded by specific end-users. In other words, they have value setting them apart from conventional agricultural commodities. Producers will receive price premiums to ensure that the valuable varieties are produced under precise agronomic regimes and segregated from non-desired varieties in the field during harvest, storage and transportation.

Beyond the farm level, the genetic modification of agricultural crops encourages the integration of agricultural production both vertically within the agricultural sector and horizontally across other sectors. Indeed, the knowledge intensification of crop development has created economic opportunities by opening up new customers for agricultural production, such as 'pharming' or industrial speciality oils.

Along with the private economic benefits, there are public economic benefits made possible by GM-crop development. Consider first that GM crops are poised in the short to medium term to alleviate the demands on domestic farm-support programmes. As previously discussed in Chapter 2, GM crops essentially represent scientific solutions to the public policy problems of variance in agricultural crop quantity and quality. Endogenous technological innovation embedded in the seed can improve crop performance and decrease the reliance of the agricultural sector upon domestic support programmes. In the long term, GM crops are poised to become a major foundation of a nation's industrial production base, as applications are found across non-food industries. In fact, GM crops must be understood as a competitiveness issue of national economic importance with significant global implications. GM-crop technologies allow for the creation of endogenous comparative advantage (Grossman and Helpman, 1991), whereby comparative advantage is no longer based on the natural endowment of factors of production. Instead, the technologies allow countries to shape their own comparative advantage and determine their economic future. Attracted by this potential outcome, public policies have been designed to ensure an internationally competitive agricultural biotechnology capacity during the so-called agriculturalization of industry (Shimoda, 1997).

Given the potential economic benefits of GM crops, it is easy to understand support among some interests for the research, development and commercialization of these varieties. Yet, in order to take advantage of the private and public economic opportunities of GM crops, the jurisdiction must have capacity. Capacity represents the ability of both private and public personnel to apply modern biotechnology. This requires a sophisticated national infrastructure in terms of human capital, capital equipment and investment capital. Therefore, it is not difficult to see why domestic governments in pursuit of the economic potential have supported the capacity building of biotechnology in the agricultural sector through various public policy initiatives.

Building capacity is, however, a formidable challenge, where capacity is a function of both time and space. In respect of time, it requires years to acquire the level of scientific human capital necessary to conduct research and development at the frontiers of agricultural biotechnology. According to Baltimore (1982), the acquisition of the necessary human capital alone requires 30 years of educational development per researcher. In addition, since most of the current initiatives in agricultural biotechnology are driven by the private sector, capacity also requires a mature financial system capable of channelling investment capital towards specific agricultural biotechnology applications.

In respect of space, capacity is linked to the existence of a critical mass of research activity. As previously discussed in Chapter 2, advancements in modern biotechnology generally are applicable across more specific applications, so that a critical mass of research activity focused broadly on advancements in biotechnology assists the development of capacity (Barton, 1998; Theodorakopoulou and Kalaitzandonakes, 1999). It is argued that such a research capacity must have a geographical profile (Zilberman *et al.*, 1997; Phillips and Khachatourians, 2001). Even between developed countries, the private and public economic importance of GM crops creates economic competition on the frontiers of science, where the concentration of capacity may result in the formation of 'agricultural industrial complexes' in the developed countries (Shimoda, 1997; Zilberman *et al.*, 1997).

The capacity issue of agricultural biotechnology has dynamic economic scale effects. A critical mass of research activity increases the potential for the development of new GM products, while learning effects decrease the time required for GM-product development (Barton, 1998).

An important aspect of the capacity issue in developed countries is the fact that it has been driven by the private sector. As the state has retreated from both basic and applied research into the applications of agricultural biotechnology, the void has been filled by private firms. This shift in leadership has been readily supported by governments in North America and Europe. Indeed, Table 3.1 provides evidence as to the 'private' capacity that exists in GM-crop development.

Following this economic rationale, the supply side of the agricultural sector has undergone structural change in order to build capacity. Increasingly, it has abandoned its traditional structure, where agricultural products were bulk commodities and producers were considered to be a homogeneous group. Very little integration occurred as commodities moved along the supply chain through spot market transactions. As

Table 3.1. Examples of integration in the agricultural sector.

Monsanto	1997 acquires Calgene (bio and seed) 1997 acquires Agracetus (bio) 1997 acquires Asgrow Agronomics (seed) 1998 (May) strategic relationship in DelKab Genetics (bio and seed) 1998 (May) ownership position in Delta and Pine Land (seed); cancelled in January 2000 by Monsanto 1998 (May) joint venture with Cargill for food processing and packaging 1998 acquires Holden's Foundation Seeds (seed) $1100 million 1999 merger with Pharmacia & UpJohn, where the agricultural biotechnology division would be made into a separate legal entity apart from the pharmaceutical operations
Dow Elanco	1997 acquires majority in Mycogen (bio and seed)
Hoechst	1997 in agribusiness venture with Agrevo and Schering acquires majority in Plant Genetic Systems (PGS) (bio) $730 million Proposed merger with Rhone-Poulenc 'Aventis', where agricultural biotechnology (Aventis Agriculture) is separated from pharmaceutical biotechnology (Aventis Pharma)
DuPont	1997 strategic partnership with Pioneer Hi-Bred (PHB) 20% stake establish 'Optimum Quality Grains' 1999 (March) acquires remaining stake in PHB $7700 million
Novartis	Merger of Sandoz (pharma), Ciba-Geigy, Ciba Seeds, Northrup-King Seeds 1999 merger with Astra-Zeneca 'Syngenta AG' to take over the combined agricultural biotechnology
Astra-Zeneca	1999 merger of Astra (pharma), Zeneca £53,000 million 1999 merger with Novartis 'Syngenta AG' to take over the combined agricultural biotechnology

mentioned, production-trait GM crops remain largely congruent with this traditional system, which is in part responsible for their popularity.

However, output-trait applications or bioengineered products require a greater level of vertical integration along the supply chain to ensure segregation. End-users contract for the production of customized GM varieties tailored to meet their specific demands. These GM varieties, grown for food or non-food use (e.g. pharmaceutical or industrial chemical uses), must be segregated from other varieties to ensure value capture – that is, to ensure the production and delivery of the high-value GM varieties that end-users want. Additionally, non-food-use GM crops must be segregated for safety reasons to ensure that such GM-crop varieties are kept out of the food supply, because they have not been designed for food production. The agricultural supply chain must become vertically integrated in order to facilitate the research, development, production, distribution, marketing and end use of customized, differentiated GM agricultural products.

Recent restructuring in the traditional agrochemical industries supports this and is indicative of the economic need for vertical integration (see Table 3.1 for some examples). Much economic analysis has focused on the causes of vertical integration, where the key cause is the knowledge intensification of the crop development process. In fact, the knowledge intensification of the GM-seed industry has created a practical need to ensure segregation for value capture and a commercial need to protect advanced knowledge. This in turn has created powerful economic incentives for firms in the agricultural life-sciences sector to engage in significant vertical and horizontal integration. It is argued that the result of this integration is the 'industrialization of agriculture' or the 'agriculturalization' of the national economy (Shimoda, 1997; Zilberman *et al.*, 1997; Barton, 1998).

As knowledge, embedded in the seed, has become the future of agricultural production, GM varieties will break the dependence of intensive agricultural production on chemicals. Traditionally, chemicals were added postseeding to overcome the production risks, but with modern biotechnology the ability to overcome production risks can be bred into the seed itself. Realizing this, agrochemical firms have moved upstream to acquire the seed and biotechnology firms and hence to own the knowledge. In the short run, production-improved herbicide-tolerant GM varieties have been developed as part of an integrated seed–chemical regime, supported by the agrochemical firms.[1] This shifts the chemical use from many different chemicals to one non-selective, broad-spectrum herbicide, hence decreasing the commercial opportunities of agrochemical companies.

[1]For instance, Monsanto's Round-up-Ready herbicide-tolerant GM varieties (cotton, maize, soybean and canola/rape-seed) and Agrevo's Liberty Link herbicide-tolerant GM varieties (e.g. canola/rape-seed, maize, soybean).

But this is a short-run situation. The long run is better characterized as chemical-free production, due to the endogenous innovation within the seed. For instance, GM crops with Bt properties do not just shift chemical use, they actually decrease overall chemical use. Consequently, the agrochemical firms have moved upstream to buy into the knowledge base of the future, rather than remaining dependent on chemicals, which are an ever-decreasing share of production inputs.

Economists argue that this upstream movement is both predictable and consistent with economic theory (Kalaitzandonakes and Hayenga, 1999). The agrochemical firms seek to identify, patent and protect the industrial base of their future – knowledge of plant genomics (Joly and Lemarie, 1998). But this is no different from the multinational pharmaceutical giants, which have long been accepted and tolerated. Similar to pharmaceutical companies, agricultural 'life-sciences' firms also face significant research and development (R&D) costs for GM crops and capacity is made more efficient through economies of scale. Two economic studies have concluded that vertical integration is driven by an economically rational and efficient response to the need to identify, patent and protect the knowledge intensification of the seed industry (Graff *et al.*, 1999; Rausser *et al.*, 1999). This conclusion, combined with the conclusion that agrochemical firms have moved upstream because of the limited future of their chemicals, challenges the popular argument among many critics of GM crops (see Chapter 4) that life-sciences firms are monopolists in pursuit of anticompetitive market power in order to increase the dependence of agricultural production on chemicals. On the contrary, vertical integration is driven by both the knowledge intensification of the seed industry and the subsequent need to build capacity and secure a competitive position in the agricultural sector by increasing a stake in biotechnology and decreasing the reliance on synthetic chemicals alone.

The firms developing GM crops face two important issues associated with the knowledge intensification of agricultural crops. First is how to be compensated for the knowledge embedded in the seed. One approach is the legal approach. Seed purchases have shifted from traditional spot transactions for conventional seeds to complicated technology-use agreements (TUAs) for GM seeds, associated with a premium 'technology fee'. In 1999, Monsanto charged a technology fee of US$6.50 bag^{-1} of Round-up-Ready soybean seeds, which represented a 40% premium on the GM seeds over non-GM seeds. Further, TUAs for Monsanto's herbicide-tolerant soybeans prohibit both the resale of the seeds by the producer (prior to planting and postharvest) and the practice of seed-saving (using harvested seeds for the next crop). While the technology fee compensates the GM-seed developer at the time of purchase, the TUAs attempt to ensure that the developer is compensated for the next crop as well.

An alternative approach for ensuring compensation for the commercial use of the knowledge embedded in the seed is the biological approach, through the use of so-called 'terminator' or 'traitor' technology.[2] The objective is to develop GM seeds that are sterile and cannot be used for a second planting or cannot cross-pollinate with other plants. Simply, this approach bypasses the need for legal TUAs. This is actually not a new approach, as F_1 hybrid varieties of maize seeds used for decades are sterile after harvest and cannot be successfully planted the following crop season.

The second important issue for GM-crop developers, associated with the knowledge intensification of the seed industry, is how to protect their knowledge from other firms. A common approach to vertical integration has been through acquisition and consolidation, rather than other tools of industrial organization such as licensing agreements or contractual arrangements for technology use between biotechnology, seed and agrochemical firms. It has been argued that acquisition and consolidation have been the chosen strategy, again not for anticompetitive monopoly reasons, but because of the current regime for intellectual-property protection (Rausser *et al.*, 1999). Essentially, outright ownership of the knowledge is the sure way to protect it. If intellectual-property protection laws were stronger and internationally respected, then technology licensing would be the chosen strategy and not direct ownership. The economic rationale behind intellectual-property rights (IPRs), such as patents and plant-breeders' rights (PBRs), is that they provide an incentive both for investment in inventive activities and for technology transfer (Malchup, 1958). Zilberman (1999) argued that a weak IPR regime over GM-crop technologies has encouraged too much agroindustrial concentration, because only direct ownership will ensure that intellectual property is protected.

The horizontal integration of agricultural production simply represents an interindustry extension of the issues associated with the vertical integration of the agricultural sector. Horizontal integration is motivated by the output-trait applications and bioengineered products, allowing GM crops to meet the demands of new end-users, such as pharmaceutical, nutriceutical or industrial firms.

Given the fundamental role of capacity in the R&D and commercialization of GM crops, the significant public and private investment in capital that must be made in order to build capacity and the necessary restructuring of the agricultural sector, economic interests naturally support a stable regulatory framework that encourages technological progress through a scientific-rationality approach to risk

[2]Terminator technology is patented by the US Department of Agriculture (USDA) and Delta and Pine Land, now owned by Monsanto, while traitor technology was being developed by UK-based Astra-Zeneca. However, this technology is not in use.

analysis and that clarifies IPRs. This is not to suggest that economic interests ignore safety issues. On the contrary, economic interests generally provide significant scope for safety issues in regulatory frameworks. They often insist, however, that there is a sound scientific basis for determining safety, subject to an objective, rules-based analysis of risk, so that stability and commercial predictability are built into the framework.

3.2 Regulatory Integration: the Traditional Trade Approach

The objective of this section is to examine what type of regulatory integration strategy is congruent with the regulatory approach supported by economic interests. It will be argued that the international economic-integration approach has led to the current international trade regime characterized by the World Trade Organization (WTO), its agreements and its affiliated international institutions.

International economic integration and regulatory competition have dominated the traditional trade-diplomacy approach to dealing with regulations. This approach is characterized as closed-door, non-transparent market-access negotiations, resulting in political compromises and concessions between sovereign states. The negotiations are aimed at establishing trade agreements outlining certain and predictable rules for the market access of traded products. They have been built on the premise that industrial interests lobby domestic governments to provide commercial protectionism from foreign imports or support for domestic exporters. Trade agreements attempt to limit the ability of governments to acquiesce to this pressure in order to enhance international economic integration (Perdikis *et al.*, 1999).

Further, the traditional trade approach attempts to disentangle trade barriers erected because of safety reasons from those erected for non-safety reasons. The former are subject to a scientific justification for the safety measure. In the event of a justification, it is legitimate for a country to impose a unilateral safety barrier to particular imported products. The latter – non-safety measures – are subject to the traditional trade principles of non-discrimination. In the event that a country imposes a trade barrier against a certain product, this barrier must be equally enforced across all similar or 'like' products, both domestic and foreign.

Social regulatory barriers facing GM crops have emerged as a contentious issue facing the international trading regime.[3] For instance, at the 1999 Ministerial Meeting of the WTO in Seattle, Canada and the

[3]For a history of the trading regime see Grimwade (1996) and Jovanovic (1998e).

USA jointly proposed a World Trade Biotechnology Initiative under the auspices of the WTO. There were two objectives: first, to establish an international fact-finding group to examine the trade issues raised by the development and commercialization of GM crops and, secondly, to establish binding international trade rules for GM crops. While the European Union (EU) accepted the first objective, it completely rejected the second objective, stating:

> We reject requests to deal with biotechnology exclusively on trade grounds. We reject market access negotiations for GMOs [genetically modified organisms]. We reject any attempt to undermine the EU right to regulate. And we reject any attempt to derail, divert or delay the biosafety talks.
>
> (*AgraFood Biotech*, 1999c)

This statement clearly illustrates the trade controversy associated with GM crops and social regulatory barriers. Basically, the controversy arises because some countries are in pursuit of an international economic-integration approach to GM-crop regulations (i.e. Canada and the USA), which is largely the approach of traditional trade diplomacy, while other countries are in pursuit of an international social-integration approach to GM crops, as evidenced by the EU support for the Cartagena Protocol on Biosafety (see Chapter 4). Moreover, due to the failure to establish the World Trade Biotechnology Initiative, there is no coordinated WTO initiative for addressing trade issues associated with GM crops.

The question may be raised: what is it about the traditional trade-diplomacy approach that the EU rejects? To understand the traditional trade approach to regulatory integration supported by economic interests, it is necessary to examine how this approach tends to deal with social regulations, such as food safety and environmental protection. Essentially, the objective is to identify the 'trade grounds' that the EU has rejected for assessing social regulatory barriers.

There is common ground between food-safety and environmental-protection social regulations. First, both are traditionally domestic areas that have qualified for exemptions under international trade agreements, so that their relationship with the rights and obligations specified in the trade agreements remains controversial and uncertain. Secondly, domestic regulations to ensure food safety and environmental protection employ precaution in the face of uncertainty when scientific evidence is insufficient – the so-called precautionary principle (discussed in more detail in Chapter 5). Thirdly, the demand for food-safety and environmental-protection measures is social protectionism, generally emerging from non-industrial actors, such as consumer and environmental organizations. Fourthly, the demand for both food-safety and environmental-protection measures is income-elastic – that is, countries with higher incomes have higher standards for food safety and environmental protection (Wang and Caswell, 1997). In short, both food-safety and environmental-protection measures can be social regulatory barriers. In fact, it is often difficult to disentangle GM-crop regulations to ensure food safety from those designed to ensure environmental protection. Together, these similarities lead to the principal trade concern with food-safety and environmental-type social regulatory barriers; while they incur market-access delay and prohibition, there are uncertain trade rules outlining their legitimate use. It will be argued that the international trade regime – driven by economic interests – has developed a two-prong approach for dealing with both food-safety and environmental-protection measures. The approach is to separate safety measures from non-safety measures, where the former are subject to scientific justifications of risk, while the latter are subject to traditional trade principles of non-discrimination.

Despite the similarities, however, currently under the WTO there is greater discipline on food-safety-type social regulatory barriers than those for environmental protection. Yet these food-safety rules have been controversial and, in the case of hormone-treated beef, have created a significant transatlantic trade-diplomacy challenge. Indeed, it is worth considering the threat that as the WTO reinforces its rules for food safety, anti-GM-crop pressures will increasingly shift from food-safety to environmental-protection and social-norm justifications for social regulatory barriers – regulations that are less rules-based and less science-based, making them more difficult to discipline and hence to integrate.

3.2.1 Food safety and trade

Both the international trade of agricultural products and the domestic regulation of food have a long history. Trade objectives of market liberalization and the removal of market fragmentation have not always been congruent with domestic regulations designed to ensure the safety, wholesomeness and adequate labelling of food products. The purpose of this section is to examine how food-safety issues are dealt with according to the traditional trade-diplomacy approach of the WTO.

It is difficult to separate food-safety issues from those associated with food quality. Some have argued that safety is really only one of several attributes that together contribute to the quality of a food product (Caswell and Hooker, 1996). Therefore, although this section is focused on food safety, a discussion of food quality will at times be necessary.

In general, food-safety issues are dealt with at the WTO under the Agreement on Sanitary and Phytosanitary Measures (SPS Agreement) while non-safety food-quality issues are dealt with at the WTO under the Agreement on Technical Barriers to Trade (TBT Agreement). Both agreements defer to standards developed in international organizations. For instance, the SPS Agreement defers to the international measures on food safety established under the Codex Alimentarius and the International Plant Protection Convention (IPPC), two international institutions which will also be examined in their trade context. In fact, the IPPC provides an illustrative example not only of food-safety measures established at the international level, but also of the difficulty of disentangling food-safety measures from environmental-protection measures.

A. The Agreement on Sanitary and Phytosanitary Measures

The SPS Agreement was the product of a convergence of economic interests, such as food-exporting countries and multinational food-processing and distributing companies, who shared a common concern about the market-access barriers facing the food trade. They believed that previous trade agreements simply did not provide enough discipline to protect their international economic-integration interests. As a result, the SPS Agreement seeks to discipline the use of food-safety measures in order to establish predictable and stable market-access rules promoting international economic integration.

By way of background, the original General Agreement on Tariffs and Trade (GATT 1948) was built on the principle of non-discrimination, summarized by the following three provisions:

1. The national treatment provision (Article I), which states that foreign products must be treated like domestic products.
2. The most-favoured-nation principle (Article III), which states that there should be no discrimination between products originating from different countries.
3. A distinction between processes and products, whereby all 'like products' were to be treated the same, regardless of the process used in their production.

These principles essentially mean that 'like' or 'substantially equivalent' products must be subject to the same regulations in a particular regulatory jurisdiction, regardless of their origin or the production and processing methods (PPMs) used in their production.

The GATT 1948 was, however, unclear on the issue of trade rules and domestic food-safety measures. Signatory countries retained significant discretion to establish their own food-safety and food-quality regulations according to a number of the GATT articles. For instance, Article XI specifically permitted regulations setting out national 'standards or regulations for the classification, grading or marketing of commodities in international trade'. Article XX(b) permitted the adoption or enforcement of measures necessary to protect human, animal or plant life or health. In an attempt to be consistent with the principle of non-discrimination, the discretionary measures invoked under Articles XI and XX(b) were not to be applied in such a manner as to cause arbitrary or unjustifiable discrimination between countries or disguised restrictions on trade. However, this specification did not provide any discipline on the type of measure that could be implemented.

The SPS Agreement, driven by economic interests, attempted to deal with the contentious ambiguities associated with discretionary food-safety standards by specifically outlining permissible types of trade-restricting measures that WTO members may enact in order to protect human, animal and plant safety and health from the import of agricultural products. In this sense, it is consistent with an economic approach to disentangling legitimate social regulatory barriers from illegitimate ones.

Accordingly, the objective of the SPS Agreement is to outline rules for when and how members can deny market access to particular exporters because of the risk that imports will contain pests or diseases. Unsafe imports can jeopardize human safety and health either directly, by making imported foodstuffs unsafe, or indirectly, by infecting domestic food inputs, including livestock and agricultural plants that are part of the domestic food-chain. There is a crucial distinction to note. The SPS Agreement targets measures taken to protect the domestic food supply, not measures taken to target overall domestic biodiversity. In this sense, the SPS Agreement relates to food-safety measures only, not environmental-protection measures, although in practice this distinction is blurred.

The SPS Agreement states that 'no member should be prevented from adopting or enforcing measures necessary to protect human, animal or plant life or health'[4] which remains in accordance with the traditional exemption provisions under Article XX(b) of GATT (1994). In the SPS Agreement, the risks that measures may target are those arising from:

- the entry, establishment or spread of pests, diseases, disease-carrying organisms or disease-causing organisms;
- additives, contaminants, toxins or disease-causing organisms in food, beverages or feedstuffs;
- diseases carried by animals, plants or products thereof.

(SPS Agreement, Annex A)

To prevent imported products from jeopardizing the safety of the domestic food supply, members may restrict or prevent imports through the use of mandatory sanitary and phytosanitary measures.

[4]Agreement on the Application of Sanitary and Phytosanitary Measures, Uruguay Round of Multilateral Trade Negotiations Legal Texts, pp. 69–84 (the 'SPS Agreement'), Preamble.

There are four important provisions of the SPS Agreement that differ from traditional trade principles and support the unilateral establishment of SPS measures by members.

First, under the SPS Agreement, members may discriminate against imports because of the presence of the above risks in the exporting country (SPS Agreement, Article 2:3). The agreement recognizes that different regions with different geographical conditions and agronomic practices face different incidence of pests and disease. As a result, it is not possible to establish uniform SPS measures to apply to all exporters, according to the principles of non-discrimination. Instead, trade measures need to specifically target those imports that may contaminate the domestic food supply, while other imported agricultural products may not face the same measures. This provision is an important exemption to the traditional non-discrimination principles. Members are not required to grant either national treatment or most-favoured-nation (MFN) status to agricultural exporters whose products risk contaminating the domestic food supply.

Secondly, according to the agreement, members may also establish domestic SPS measures higher than the accepted international standard if there is scientific justification for doing so (SPS Agreement, Article 3:3). Generally, international trade agreements commit members to adopting international standards if available; however, the SPS Agreement permits members to establish even higher standards.

Thirdly, the SPS Agreement permits members to establish SPS measures based on scientific risk, as well as broader assessments of risk, such as relevant economic factors, which include:

- the potential damage in terms of loss of production or sales in the event of the entry, establishment or spread of the disease or pest;
- the costs of control or eradication in the territory of the importing Member;
- the relative cost-effectiveness of alternative approaches to limiting risks.

(SPS Agreement, Article 5:3)

Trade agreements traditionally avoid such socio-economic assessments because of the subjectivity complications that are associated with them. Indeed, as previously discussed, the economic perspective attempts to depoliticize trade and make it a function of comparative advantage (WTO, 1995), and yet the SPS Agreement recognizes the socio-economic nature of food-safety regulations and permits such consideration.

Fourthly, and finally, under the SPS Agreement, members may establish provisional SPS measures based on precaution, in the event that there is insufficient scientific evidence to conduct an appropriate risk assessment. The Agreement states:

> In cases where the relevant scientific evidence is insufficient, a Member may provisionally adopt sanitary or phyto-sanitary measures on the basis of available pertinent information, including that from sanitary or phyto-sanitary measures applied by other Members. In such circumstances, Members shall seek to obtain additional information necessary for a more objective assessment of risk and review the sanitary or phyto-sanitary measure accordingly within a reasonable amount of time.
>
> (SPS Agreement, Article 5:7)

That is, members are permitted to establish trade barriers based on the precautionary principle. These barriers can remain in place until enough scientific evidence about the risk has been compiled.

While the SPS Agreement allows members considerable scope to impose unilateral social regulatory barriers, there are important conditions on the sanitary and phytosanitary measures that members may unilaterally impose in order to prevent them from being used as disguised protectionist barriers. It will be argued that these conditions reflect the predominant influence of the scientific-rationality perspective on regulatory development and regulatory integration supported by economic interests.

To facilitate international trade, members are committed to both the international harmonization of SPS measures and to the mutual recognition of measures employed by other members. In respect of harmonization, members commit to adopting international standards, guidelines or recommendations as the prevailing national standards, in order to promote international harmonization (SPS Agreement, Preamble). For food safety, the relevant international institution is the Codex Alimentarius Commission (CAC), for animal safety the International Office of Epizootics (OIE) and for plant safety the IPPC. In respect of mutual recognition, members are committed, in principle, to granting equivalence to the SPS measures adopted by exporting countries 'if the exporting Member objectively demonstrates to the importing Member that its measures achieve the importing Member's appropriate level of sanitary or phytosanitary protection.' (SPS Agreement, Article 4:1). To facilitate the process, the importing member must be allowed to conduct a conformance assessment, including inspection, testing, monitoring and evaluation of the measures in place in the exporting member.

Yet, as mentioned, members are permitted to exceed international standards provided there is a scientific justification for doing so. Hence, an important condition is a scientific justification. According to the Agreement, unilateral SPS measures must be 'based on scientific principles' and cannot be maintained 'without sufficient scientific evidence' unless it is a temporary, precautionary measure (SPS Agreement, Article 2:2). The science-based measures adopted must be proportional to the risk that is being targeted. In order to assess the risks, members are committed to considering the risk-assessment techniques used in the international standard-setting institutions, even if the relevant international standard is not being used (SPS Agreement, Article 5:1). Further, the Agreement states that:

> In the assessment of risks, Members shall take into account available scientific evidence; relevant processes and production methods; relevant inspection, sampling and testing methods; prevalence of specific diseases or pests; existence of pest- or disease-free areas; relevant ecological and environmental conditions and quarantine and other treatment.
>
> (SPS Agreement, Article 5:2)

Hence, the SPS Agreement requires members to provide a scientific justification for the adoption of measures, where the scientific justification is crucial in supporting the domestic measure in the event of a trade challenge.

> When a Member has reason to believe that a specific sanitary or phytosanitary measure introduced or maintained by another Member is constraining, or has the potential to constrain, its exports and the measure is not based on the relevant international standards, guidelines or recommendations, or such standards, guidelines or recommendations do not exist, an explanation of the reasons for such sanitary or phytosanitary measure may be requested and shall be provided by the Member maintaining the measure.
>
> (SPS Agreement, Article 5:8)

According to the Agreement, in the event of a trade dispute over the use of a food-safety measure, the WTO Dispute Settlement Panel seeks the scientific advice of the CAC. Without an acceptable scientific justification, it is unlikely that a trade-dispute decision by a WTO dispute-settlement panel or an appellate body will support the unilateral SPS measure. In this sense, even if members do not adopt international standards, it is important that domestic food-safety measures remain congruent with the international risk-analysis approach of the CAC in the event of a trade dispute. The CAC will be discussed in more detail in subsection C below.

The SPS Agreement also commits members to publishing a draft of the domestic measures to the SPS Committee and allowing for a 60-day review and comment period for all concerned exporters (SPS Agreement, Annex B:5). The logic is that such a review and consultation process may proactively avoid future trade disputes by ensuring that the food-safety measures adopted by a member take into account the process and production realities in exporting countries. The Agreement obliges, although does not require, the importing member to take full account of the comments and endeavour to ensure that the SPS measure fulfils a legitimate and scientifically justifiable safety objective without unduly affecting agricultural trade.

The Marrakech Agreement 1994 included a provision for the SPS Committee to conduct a review of the Agreement in order to examine its progress in enhancing trade liberalization and to identify points of clarification that may be addressed through amendments. The general interpretation of the 1998 review was that nothing could be amended without creating significant controversy and/or requiring an opening up of the Agreement. Essentially, the SPS Agreement has been established to be congruent with a scientific-rationality approach to economic integration, and decision-making has favoured economic trade interests. Dealing with concerns would require an opening up of the Agreement to allow for greater influence from the social-rationality perspective. As a result, the Agreement was left virtually unchanged.

The SPS Agreement aims to establish predictable and certain rules for legitimate social regulations on a sound scientific basis. It represents a 'safe harbour' for members, since it allows them to take unilateral trade-restriction measures against agricultural imports to address domestic concerns about food safety in the face of risk and uncertainty, subject to conditions on the unilateral use of SPS measures in order to prevent unnecessary trade distortions. The challenge is finding the line between SPS measures that legitimately restrict trade in order to protect human, animal and plant safety or health and those SPS measures that unnecessarily restrict trade. According to the Agreement, the most objective way to define the line is through a scientific risk-assessment procedure according to the risk-analysis framework supported by the CAC. An SPS measure used by a member will be considered legitimate if there is sufficient scientific proof for its use and if it is the measure with the lowest cost to the consumer and the international trading system (Roberts, 1998).

Indeed, this approach should come as no surprise, as the SPS Agreement came about because it was in the economic interests of large, multinational agrifood companies and the economic interests of export-orientated countries. During the Uruguay Round negotiations, these interests held significant policy power, even in the EU. Phillips (1991) argued that the EU approached these negotiations from the economic perspective and with an export orientation. A broad coalition of interests, including farmers and consumers, viewed the SPS Agreement as a potential win–win situation, whereby market-access rules could be clarified and yet social food-safety regulations would be protected. Hence, the EU was in support of the Agreement's international trade rules for food-safety-type social regulatory barriers.

With respect to the use of biotechnology in food production, the SPS Committee has deferred to the work of the CAC in establishing both legitimate scientific risk analysis procedures and regulatory guidelines based on those procedures. Hence, it is crucial to examine the CAC below.

In summary, it is important to note at this point that, according to the SPS Agreement, food-safety-type social regulatory barriers to GM crops will be assessed in terms of their scientific justification and their trade impact. In practice, this means that there is no scope under the WTO to impose legitimate trade barriers predicated on social preferences beyond a scientifically justified prevention of risk. In addition, social regulatory barriers that are justified must also meet the requirement of minimum trade disruption. In other words, trade and market-access imperatives dominate even justified social regulatory barriers according to the economic-integration approach of the SPS Agreement.

B. The Agreement on Technical Barriers to Trade

The TBT Agreement deals with technical, non-safety, food-quality issues, such as nutrition analysis, grading, packaging (labelling, symbols, markings, terminology) and protection against deceptive or fraudulent practices. The TBT Agreement also has provisions on establishing conformity assessments between trading partners. Conformity assessments, which are important for mutual recognition efforts, are any procedures used, either directly or indirectly, to determine that relevant requirements in technical regulations (mandatory requirements) and standards (voluntary requirements) are fulfilled, which include: sampling; testing and inspection; evaluation, verification and assurance; registration, accreditation and approval. Further, the TBT Agreement deals with broader issues of food trade not easily captured in a science-based framework of risk analysis. Therefore, the role of the TBT Agreement in the food trade is important and needs to be assessed.[5]

Historically, concern about the potential impact of technical standards and product labelling upon international agricultural trade has been voiced at the multilateral level since the mid-1970s, driven mainly by economic interests in Australia, Canada and the USA. The primary concern was that technical food standards, including labelling schemes, might be used either to restrict market access or to confer an advantage on domestic products in the domestic marketplace. Yet such standards could be used in a discretionary manner with no real discipline on their application.

The Tokyo Round of the GATT introduced international trade discipline of food-quality regulations, including labelling measures, through the TBT Code. The essence of the TBT Code was to establish international obligations on technical regulations, standards and con-

[5]The TBT Agreement also has relevance to environmental-protection issues associated with agricultural biotechnology, to be discussed in section 3.2.2.A. The relevance of the TBT Agreement to environmental protection lies in its jurisdiction over PPMs.

formance assessments for both transparency and notification, based on the trade principle of non-discrimination (OECD, 1997c; Caldwell, 1998).

Yet, despite this Code, agricultural exporters still had concerns with food-quality measures, because such measures were generally still exempt from any discipline. They wanted to see the TBT Code, which only applied to a limited number of developed contracting parties, strengthened and extended to cover all contracting parties. As a result, negotiations in the Uruguay Round produced the TBT Agreement.

There are three similarities between the SPS and the TBT Agreements. First, the TBT Agreement allows members to establish trade-restricting measures in order to protect human and environmental health and safety and to ensure the quality of imported products, the so-called 'legitimate objectives', provided that the measures do not unnecessarily obstruct international trade (TBT Agreement, Article 2:2) – that is, similarly to the SPS Agreement, the social regulations of members may be trade barriers to imported products.

Secondly, it requires members to base their national standards on international measures established by international standard-setting bodies (TBT Agreement, Article 2:4). When internationally agreed standards cannot be adopted due to geographical, climatic or technological reasons, the member must publish the draft measures in order to allow potentially affected foreign producers an opportunity to respond to them (TBT Agreement, Article 2:9). It is anticipated that concerns of exporters will then be incorporated into any subsequent measures.

Thirdly, the TBT Agreement requires that, where applicable, national measures should be scientifically justifiable (TBT Agreement, Article 2:2). The Agreement includes specified criteria that members must account for in formulating TBT measures in order to ensure that measures do not create unnecessary regulatory barriers to trade. Similarly to the SPS Agreement, the TBT Agreement employs a scientific-rationality approach in order to identify legitimate social regulatory barriers.

Despite the similarities, there are four differences between the SPS and the TBT Agreements. First, unlike the SPS Agreement, which permits discrimination in the application of trade-restricting measures, the TBT Agreement is based on the traditional trade principles of non-discrimination. The Agreement states that measures should be applied on an MFN basis to all imported products from all contracting parties (TBT Agreement, Article 2:1 – MFN Principle of Non-Discrimination). It also states that measures should not extend to imported products treatment that is less favourable than that extended to domestically produced 'like' products (TBT Agreement, Article 2:1 – National Treatment Principle).

Secondly, whereas the SPS Agreement deals with mandatory national food-safety measures, the TBT Agreement deals with both mandatory (technical requirements) and voluntary (standards) measures. Both mandatory and voluntary measures can address product characteristics, PPMs, terminology and symbols and packaging and labelling requirements (i.e. prevention of deceptive advertising practices). Voluntary standards are subject to the TBT Code of Good Practice for the Preparation, Adoption and Application of Standards, which urges members to make their best endeavours to ensure that voluntary trade-restricting measures are subject to the same principles and rules as mandatory standards (TBT Agreement, Article 4 and Annex 3). The Code urges members to use international standards as a basis for national voluntary standards and to participate fully in the preparation of international standards.

Thirdly, unlike the SPS Agreement, the TBT Agreement does not allow for provisional trade restriction based on precaution where scientific evidence is insufficient.

Fourthly, although the TBT Agreement requires that measures should be scientifically justifiable, the problems with determining appropriate scientific risk-assessment procedures for non-safety issues and other legitimate objectives, such as labelling for the consumers' right to know, are enormous. As a result, the scientific-justification principle under the TBT Agreement is considerably weaker than that under the SPS Agreement, potentially permitting food-quality-type social regulatory barriers. For instance, under the TBT Agreement, labelling standards can prevent deceptive marketing practices that have an adverse impact on informed consumerism, but, since scientifically justifying adverse impacts on informed consumerism is difficult, such measures do not really require a science basis for justification. Specifically, in respect of GM crops, the TBT Agreement's mandatory and voluntary labelling provisions are most relevant.

The WTO's Committee on Technical Barriers to Trade provides a forum for members to raise concerns about the labelling legislation of other members and how this legislation might contravene the rights and obligations under the agreement. A recent example is a US submission to the TBT Committee underlining concerns with the EU labelling regulation 1139/98 (the GM Soya/Maize Regulation) (WTO, 1998d). The US submission indicated that the USA felt that bilateral consultations with the EU on the proposed regulation did not address fundamental US concerns, supported by Brazil, Canada and New Zealand. The submission states that 'the United States urgently requests that the EC address our concerns and comply with its obligations under the Agreement, in respect of these and future regulations.' Essentially, the US position was that Regulation 1139/98 could not achieve its alleged 'legitimate objective' and that the implementation rules make compliance by producers from other members overly difficult, concluding that it is an unjustified barrier to trade. Labelling-based trade disputes at the WTO are discussed more fully in Chapter 5.

C. The Codex Alimentarius

Economic interests support the work of the Codex in establishing international food-safety and food-quality measures, because the Codex approach has always been congruent with the scientific-rationality approach to regulatory development. And, due to its focus on scientific rationality, economic interests supported the linking of the Codex with international trade agreements. For instance, the formal linking of trade rules and Codex food standards was examined in 1991 at the Food and Agriculture Organization/World Health Organization (FAO/WHO) Conference on Food Standards, Chemicals in Food and Food Trade held in cooperation with the GATT Secretariat. The motivation for this conference was that the draft SPS Agreement cited the Codex Alimentarius as the international institution responsible for developing harmonized food standards. Essentially, the objective of this conference was to make the Codex procedures more trade-congruent (this conference will be discussed in greater detail below).

In order to establish a science basis sufficient to justify the use of SPS- and TBT-related measures, both the SPS and TBT Agreements defer to the international food-safety and quality measures established under the Codex Alimentarius. The SPS Agreement states: 'Members shall base their sanitary or phyto-sanitary measures on international standards, guidelines or recommendations' (SPS Agreement, Article 3:1). The TBT Agreement states that:

> (w)ith a view to harmonizing technical regulations on as wide a basis as possible, Members shall play a full part ... in the preparation by appropriate international standardizing bodies of international standards for products for which they have either adopted, or expect to adopt, technical regulations.
>
> (TBT Agreement, Article 2:6)

In other words, Codex food-safety standards are relevant to the SPS Agreement, while Codex food-quality standards are relevant for the TBT Agreement.

Because of its focus on scientific rationality and its links with international trade agreements, the Codex is very vulnerable to criticisms from those who support a social-rationality regulatory approach and who reject the traditional trade approach of international economic integration. This is precisely the conflict that emerges in respect of GM crops. To better understand this conflict, the discussion begins with an examination of the scientific-rationality approach of the Codex, followed by a discussion of the current debates on GM crops and GM foods at the Codex.

The Codex Alimentarius, created in the early 1960s under the United Nations' (UN) Food Standards Programme, is a joint agency of the UN's FAO and the WHO. The FAO, established in 1945, has responsibilities covering food nutrition and international food standards, while the WHO, established in 1948, has responsibilities covering human health and food standards. The first meeting of the Joint FAO/WHO Expert Committee on Nutrition in 1950 concluded that there was a need for the international harmonization of food standards based on science. The meeting report explained that:

> food regulations in different countries are often conflicting or contradictory. Legislation governing preservation, nomenclature and acceptable food standards often varies widely from country to country. New legislation not based on scientific knowledge is often introduced, and little account may be taken of nutritional principles in formulating regulations.
>
> (FAO/WHO, 1950)

In this sense, the initial motivation for a Codex Alimentarius was to establish science-based, international standards for food safety and quality in order to enhance consumer protection and reduce market fragmentation.

In 1960, the FAO Regional Conference for Europe endorsed the idea of creating an international food code governed jointly by the FAO and WHO and based on the Codex Alimentarius Europeaus. In 1961, the Eleventh Session of the Conference of the FAO established the Statutes and Rules of Procedures of the Codex Alimentarius Commission (CAC). Also, it was proposed that the start-up costs of the CAC be supported by a trust fund, to which both national governments and industry groups made voluntary contributions (Frawley, 1987). In 1962, the Joint FAO/WHO Food Standards Conference requested that the CAC implement a joint FAO/WHO food-standards programme and create the Codex Alimentarius. Finally, in 1963 at the Sixteenth World Health Assembly of the WHO, the establishment of the Joint FAO/WHO Programme on Food Standards was approved.

The creation of the Codex Alimentarius was driven largely by desires to ensure food safety and consumer protection in the context of the rapidly growing international food trade. Widely different interests shared the demand for greater international cooperation and international leadership in food standards. For instance, on the one hand, international trade associations were frustrated by the market fragmentation of divergent regulations and hence supported international coordination efforts, while, on the other, consumers' concerns, including the food-reform movements that emerged in the postwar period, were concerned with the lack of international coordination of food-safety regulations, as the international trade of food products increased. It has been argued that:

> (t)he Codex Alimentarius was a response to a widely recognized need. It did not just happen. It was a product of a long evolutionary process involving a wide cross-section of the global community. Many people representing many interests and disciplines were involved in the process. (Frawley, 1987)

Therefore, the Codex emerged from a common desire for internationally coordinated food-safety and quality rules.

Administratively, there are three separate Codex agencies, which together work to develop the Codex Alimentarius; the CAC, the Codex Secretariat and the Codex Executive Committee. The CAC meets every 2 years. To date, there have been 23 Sessions of the CAC. Commission membership is on a country basis, where all member countries of the UN may be CAC members. Currently, the Commission has 165 member countries (representing over 97% of the world's population), who participate, in varying degrees, in the development of international food standards. Member countries are represented at CAC Sessions by national delegations, composed of senior officials (usually health officials) appointed by their governments, and these may also include industry representatives, academics and representatives of national non-governmental organizations. Although the CAC Sessions were initially the domain of the developed countries, the number of developing-country delegations has steadily increased to nearly three times the number of developed-country delegations (Frawley, 1987). There is also scope in the CAC Sessions for international non-governmental organizations (INGOs) to participate as 'observers' in order to express their points of view, with the restriction that observers cannot participate in final decision-making, only national delegations can. Most member countries have established a delegation contact point – for instance, in 1992 the UK established the National Codex Consultative Committee.

The CAC is governed according to Statutes and Rules of Procedure. The Statutes identify the reasons for the establishment of the Commission and the legal basis for its activities. For instance, according to Article 1 of the Statutes of the CAC, there are five guiding principles for the work of the Commission (CAC, 1997):

(a) to protect the health of consumers and to ensure fair practices in food trade;
(b) to promote the international coordination of all food standards among international governmental and non-governmental organizations;
(c) to establish priorities for food standards and to initiate and guide the development of draft standards along with appropriate organizations;

(d) to finalize draft standards and to publish final food standards including final standards developed by other international organizations which have been internationally accepted and to publish these standards in the Codex Alimentarius either as a regional or a world wide standard;
(e) to amend published final food standards as new evidence is collected.

The Rules of Procedure of the CAC identify how the intergovernmental body operates. The rules include the appointment of Commission officers, the frequency and operation of the Commission, procedures on voting, the establishment of subsidiary bodies and issues concerning the Commission's budget and expenditures.

The CAC establishes international measures and coordinates an international dialogue on important food-safety and quality issues through various expert committees and scientific consultations. According to the Rules of Procedure, the Commission can establish two kinds of subsidiary committees, Codex Committees and Coordinating Committees. The former, Codex Committees, are made up of 15 Commodity Committees and nine General Subject Committees (Table 3.2). Committees are chaired by a host member country and the committee may be active or dormant. Host members are very influential since they, in collaboration with the Codex Secretariat, establish the agenda of meetings and issue invitations to member delegations and observers.

Coordinating Committees have no host country because they are organized according to regions. There are five such committees representing Africa, Asia, Europe, Latin America and the Caribbean, as well as North America and the South-west Pacific. Along with the Committees, there are joint FAO/WHO expert groups, which provide advice and guidance to the Codex (Table 3.2).

The second Codex agency is the permanent Codex Secretariat, located in Rome and administered by the FAO's Food Quality and Standards Service within the Food and Nutrition Division. The purpose of the Secretariat is to provide day-to-day support for member countries as they attempt to interpret, develop and implement their national food regulations congruent with the Codex Alimentarius. The Codex Secretary is an FAO official, who serves also as the Chief of the Joint FAO/WHO Food Standards Programme.

The third Codex agency is the Codex Executive Committee (CEC). The Executive Committee meets yearly and, unlike the CAC, is organized according to principal regions: Europe; Africa; Asia; the South Pacific; Latin America and North America. Hence, the Codex Executive Committee provides the regional perspective on food safety, consumer protection and, increasingly, agrifood trade.

The Codex Alimentarius is composed of standards, codes of practice, guidelines and recommendations pertaining to food safety and quality. As of 1997, 237 Commodity Food Standards, 41 Codes of

Table 3.2. Codex Commodity and General Subject Committees and Expert Committees.

Vertical Commodity Committees	
Cereals, pulses and legumes	United States
Vegetable proteins	Canada
Tropical fresh fruits and vegetables	Mexico
Processed fruits and vegetables	United States
Fats and oils	United Kingdom
Processed meat and meat products	Denmark
Meat hygiene	New Zealand
Fish and fishery products	Norway
Milk and milk products	New Zealand
Sugars	United Kingdom
Cocoa products and chocolate	Switzerland
Edible ices	Sweden
Soups and broths	Switzerland
Natural mineral waters	Switzerland
Horizontal General Subject Committees	
Residues of veterinary drugs in food	United States
Import/export inspection and certification	Australia
Food additives and contaminants	Netherlands
General principles	France
Pesticide residues	Netherlands
Food labelling	Canada
Methods of analysis and sampling	Hungary
Food hygiene	United States
Nutrition and foods for special dietary uses	Germany
Expert Committees	
Joint Expert Committee on Food Additives (JECFA) 1955	
Joint Meeting on Pesticide Residues (JMPR) 1963	
Joint Expert Committee on Fruit Juices	
Joint Expert Committee on Quick Frozen Foods	

Hygienic or Technological Practice, 185 Pesticide Evaluations, 3274 Maximum Residue Limits for Pesticides, 25 Guidelines for Contaminants, 1005 Food Additive Evaluations and 54 Veterinary Drug Evaluations have been developed (CAC, 1997).

Since the Codex Alimentarius attempts to develop universal food-safety and consumer-protection principles based on a tradition of consensual decision-making, it should come as no surprise that the administrative process is lengthy and subject to many iterative review processes. A Codex commodity-food standard is adopted only after eight stages or steps of consultation have been completed. The eight steps are (CAC, 1997):

1. A food-safety issue is identified by a national government or a subsidiary committee of the CAC and presented at a CAC plenary session (every 2 years), where, if it is determined that a Codex food standard ought to be elaborated, the CAC or the Codex Executive Committee assigns the issue to either a Commodity or a General Subject Committee.
2. The committee presents its elaboration, based on Codex food-standard elements, to the Codex Secretariat, which produces a Proposed Draft Standard.
3. The Proposed Draft Standard is sent to all member governments and identified INGOs for review and comments.
4. Comments from step 3 are returned to the Committee who initially elaborated the food standard.
5. The committee amends the Proposed Draft Standard, subject to the review comments, and the amended Draft Standard is presented to the CAC by the Secretariat at a plenary session, where it may be adopted as a Draft Standard.
6. The adopted Draft Standard is sent to all member governments and identified INGOs for further comment.
7. Comments are returned to the Committee through the Secretariat for amendments to the Draft Standard.
8. The amended Draft Standard is presented to the CAC for adoption as a Codex Standard to be sent to member governments for acceptance.

Along with Codex commodity-food standards, the Codex Alimentarius also includes Codex General Standards, developed by the horizontal General Subject Committees, which apply to all foods.

A final Codex food standard includes several elements. First, it includes a description of the product and the essential composition and quality factors that identify the product *vis-à-vis* close substitutes. Secondly, the standard includes both identification and analysis of any additives and potential contaminants in the food product. Thirdly, the food standard incorporates established Codex requirements, such as the Codex product-hygiene requirements. Fourthly, the standard includes a complete description of the scientific procedures used to sample and analyse the product during review. Fifthly, the standard specifies any labelling requirements in accordance with the Codex General Standard for the Labelling of Prepackaged Foods.

Determination of the safety of the food product is based on a risk-analysis framework, as outlined by the Codex Committee on General Principles. Scientific risk assessment involves risk identification, characterization and exposure assessment as defined in the Redbook.[6] This includes toxicological studies of pesticide residues, microbial contaminants, chemical additives and veterinary biologicals.

[6]See Chapters 5 and 6 for a discussion of the Redbook's principles of risk analysis.

In March 1995, a joint FAO/WHO consultation proposed definitions for risk consideration activities in Codex (FAO/WHO, 1995). The conclusions of this consultation were included in the 1996 CAC Progress Report, which clarified the definitions of risk, hazard, risk analysis, risk assessment, risk management and risk communication (CAC, 1996).

At the 21st Session of the CAC (3–7 July 1995, Rome) amendments to the Codex Procedural Manual included four statements of principles concerning the role of science in the Codex decision-making process and the extent to which other factors are taken into account. The four statements of principle were:

1. The food standards, guidelines and other recommendations of Codex Alimentarius shall be based on the principle of sound scientific analysis and evidence, involving a thorough review of all relevant information, in order that standards assure the quality and safety of the food supply.
2. When elaborating and deciding on food standards Codex Alimentarius will have regard, where appropriate, to other legitimate factors relevant for the health protection of consumers and for the promotion of fair practices in food trade.
3. In this regard, it is noted that food plays an important role in furthering both of these objectives.
4. When the situation arises that the members of Codex agree on the necessary level of protection of public health but hold differing views about other considerations, members may abstain from acceptance of the relevant standard without necessarily preventing the decision by Codex.

(CAC, 1995)

Economic interests supported these amendments. They were put forward by the USA and supported by other agricultural exporters and many G77 nations, against the strong opposition of the EU and the member-state representatives in the EU delegation. The first principle was the major source of contention. It has been the traditional risk-analysis approach of the Codex Alimentarius that risk should be scientific evidence of risk to human health similar to that outlined in the SPS Agreement. The EU was attempting to broaden the risk-analysis approach to include other concerns. The first principle, however, firmly supported the traditional scientific stance of the Codex risk-analysis framework. Although the second principle mentions the consideration of other legitimate factors, it is only within the parameters of the first principle that this is possible – that is, only those other legitimate factors that enhance the health protection of consumers from identified potential hazards may be considered. Further, the third principle reinforced the linkage between the Codex and the food trade. Therefore, the amendments to the procedural manual rejected the focus of food standards on broader concerns. Finally, the fourth principle pertains to the contentious role of the Codex in risk management. Recall that risk management involves the evaluation of the risk identified in risk assessment in order to establish an appropriate regulatory response that reduces and prevents the hazard. Economic interests tend to argue that, while an internationally harmonized approach to risk assessment is an appropriate role of the Codex, the response to risk is the domain of sovereign governments, because risk management inevitably involves socio-economic issues that cannot be internationally harmonized. The pressure has been to create an opt-out situation, so that a country may abstain from the Codex food-standard decision, and the standard may still be adopted on the basis of its scientific justification. This is directly against the original intent of adopting standards by consensus and promoting the international harmonization of food standards. Also, even if a country opts out of a Codex standard, the Codex standard would be considered the prevailing international standard under the SPS Agreement and would be relevant in the event of a trade dispute. While the Codex decision-making process is preferably done by consensus among national delegations, it can be by vote in the CAC if consensus cannot be achieved on very controversial issues, such as the use of hormones in beef production.

Generally, it takes about 7 years to develop a Codex food standard (i.e. half a year for each of steps (1)–(6), while steps (7) and (8) take 2 years each). There is a fast-track procedure, which can be employed if the proposed standard is relatively uncontroversial. Under the fast-track approach, it is possible for the amended Draft Standard to be adopted at step 6 as a Codex food standard, instead of being sent for further review, if consensus has been achieved.

An important aspect of Codex standards is that they are subject to revision, as new scientific knowledge becomes available. It is the responsibility of each member country to present to the Commission and the relevant subsidiary committee new information that may require revision to a Codex standard. This information must, however, meet the Codex standards for a scientific justification.

Once a Codex standard is developed, member countries are expected to adopt the standard into national food regulations in order to promote international consumer-protection and market-access rules. According to the Codex General Principles, there are three forms of member-state acceptance: full acceptance, acceptance with minor deviations and free distribution (CAC, 1996). In this sense, the regulatory convergence strategy of Codex is one of regulatory coordination according to mutual recognition. The FAO and the WHO provide assistance for developing countries in establishing the capacity and knowledge necessary to implement food legislation congruent with the many Codex measures.

Along with standards, the Codex also develops and publishes codes of hygienic practice, codes of technological practice and guidelines and recommendations on various food-safety and quality issues. The purpose of these measures is to protect the health of consumers in areas where standards may not be practical (FAO, 1998). The codes of hygienic practice provide industry with guidance on the safe production of food products. The codes of technological practice provide industry with guidance on the safe use of process, storage and transport technology during food production. Guidelines and recommendations are more general than the codes of practice and provide industry with guidance on wider issues, such as procedures for the food trade following accidental nuclear contamination.

To assist the Commission in ensuring consumer safety and health and protecting consumers from economic fraud, consumer organizations have participated in the development of Codex measures since 1965 (FAO, 1998; see also Houston, 1987). At the 20th CAC session, it was recognized that the Commission must continue to provide an opportunity for consumer organizations to participate. However, it was reiterated that Codex decision-making is done by national delegations; therefore, it is the responsibility of the national delegations to involve consumers more effectively in the decision-making process at the national level in the various national Codex committees.

The Codex also includes, in Volume 1A General Requirements, a Code of Ethics for International Trade in Foods, established in 1979 and amended in 1995. The Code is non-binding but applies to all foods traded. This code explains that the Codex goes beyond facilitating the removal of barriers to trade. Instead, it also encourages food traders to adopt voluntary ethical practices to protect consumers and to ensure fair trade. According to the Codex General Principles:

> (i)nternational trade in food should be conducted on the principle that all consumers are entitled to safe, sound and wholesome food and to protection from unfair trade practices.
>
> (FAO, 1998)

Principally, the code is to prevent countries from exporting unsafe or poor-quality food. It prohibits dumping of foodstuffs not suitable for the domestic market into foreign markets. Under the code, there are three responsibilities of the Party of Export (FAO, 1998): first, to employ as appropriate and practicable, legal or administrative controls aimed at preventing the exportation or shipment of food that does not comply with the laws of the Party of Import or, in the absence of applicable laws, with the Codex Alimentarius; secondly, to promptly notify the Party of Import of the exportation of shipments of food found not to comply, when a legal or administrative means of preventing the exportation are not available or were unsuccessfully applied or where non-compliance was determined only after exportation; and thirdly, to make available to the Party of Import, upon request, appropriate certification, inspection and other procedures, as appropriate, with the manner of compensation for these services to be agreed between the Party of Export and the Party of Import. If an importing country finds inappropriate imports, then it can request the Party of Export to pursue the exporter according to all legal and administrative procedures in order to rectify the situation, on behalf of the importer and the Party of Import.

Essentially, Codex is a top-down approach to developing universally acceptable food standards through its elaboration and consultation procedures at the multilateral level. It reflects a scientific rationality approach to regulatory development. Once a food standard is adopted by member countries, the Codex requests and the SPS and TBT Agreements require members to incorporate the standard into any relevant domestic legislation. Under the principles of both Codex and the SPS and TBT Agreements, members retain the right to unilaterally impose more stringent food-safety regulations that may be deemed necessary to ensure domestic consumer protection. However, it is anticipated that, when countries do deviate from the Codex food standard, they do so in a scientifically justifiable manner, where scientifically justifiable is measured in respect of the Codex approach to the risk-analysis framework.

Concerns associated with linking the Codex with international trade agreements were the focus of a 1991 conference. The 1991 FAO/WHO Conference on Food Standards, Chemicals in Food and Food Trade, in cooperation with the GATT Secretariat, included several important proposed reforms to the Codex in order to make it more congruent with trade agreements. First, to accommodate the needs of trade agreements, it was proposed that the standards development process should be more rapid, including a proposal for majority voting procedures. Specifically, it was proposed that a two-thirds majority vote in favour of a standard at stage 5 would be sufficient for the adoption of a standard. This departs from the Codex tradition of consensus-based decision-making and shifts it towards judicious and timely decision-making in order to support trade interests. There was considerable support for this linkage, even among the members from the EU. Recall, from the discussion on the SPS Agreement, that economic interests that supported clarified international food-safety rules dominated the EU's food-trade policy.

Secondly, as the Codex Alimentarius is composed of standards, guidelines, codes of practice and recommendations, it was proposed that all types of Codex initiatives be considered as 'standards' under the trade agreements. This proposal was also included in the Report of the Twelfth Session of the Codex Committee on General Principles

(CCGP) in 1996 (CAC, 1996). In September 1998, the CCGP decided that, in respect of the SPS Agreement, all types of Codex initiatives are functionally the same. In April 1999, the CCGP decided that all types of initiatives were all 'food standards' according to the TBT Agreement as well. This decision was on the basis of a TBT Committee recommendation that 'governments harmonize their regulations on the basis of international standards and in the framework of Codex this applies to all the provisions which do not address the protection of consumers' health' and 'there is no difference between the various categories of Codex texts involved; for the purpose of the TBT Agreement all Codex standards and related texts correspond to the TBT definition of a standard' (Codex Secretariat, 1999).

Thirdly, a proposal to increase the transparency of Codex was also included. Traditionally, Codex had been an international institutional arrangement dominated by technical discussions among leading scientists, food-safety experts and major food industries. Prior to its linking with the SPS Agreement, there was little reason for broader interest in the work of Codex. In this sense, Codex was not a deliberately untransparent institution; instead, there was no pressure for it to be accessible to interests beyond those interested in technical food-standards issues.

The fourth proposal from the 1991 conference was to give priority to horizontal committees over vertical committees – that is, to move towards the development of general international standards or minimum requirements, rather than numerous, specific vertical standards, because the harmonization of standards across specific categories was found to be very difficult and time-consuming.

Given this institutional assessment of the Codex, it is now possible to examine, more specifically, how agricultural biotechnology and GM crops are dealt with under the Codex. Primarily, they are dealt with on a product basis, not on a technology basis. In 1995, there was an unsuccessful attempt to permanently include foods derived from GM techniques on the agenda of the horizontal Codex Committee on Special Nutritionals. There is currently neither a vertical committee nor a horizontal 'general subject' biotechnology committee. Instead, various vertical commodity and general-subject committees address issues and concerns associated with agricultural biotechnology products as they fall within traditional jurisdictions, such as the Codex Committee on Food Labelling.

The use of biotechnology in food production emerged on the Codex agenda while considerable attention was focused on the EU–US dispute over beef hormones in the CAC meetings of 1991, 1993 and 1995. Further, at the CAC meetings of 1997, the use of recombinant bovine somatotrophin (rbST) was the controversial agricultural biotechnology issue on the agenda. Indeed, an important development was that the Commission did not approve the use of rbST, on the basis that 'other legitimate factors' should be examined. With attention turned to other uses, the early considerations of GM crops appeared to be very uncontroversial. Yet, in recent years, GM crops have become an important, controversial topic of discussion and consultation in the Codex. GM crops raise two important issues for Codex. The first is determining the role of science in Codex standard-setting procedures, while the second is determining the role of Codex in risk-management procedures.

In 1990, a WHO/FAO Joint Expert Consultation examined the issue of foods produced from GM ingredients and made seven recommendations:

1. GM (rDNA) [ribosomal DNA] foods should be evaluated for both safety and nutritional value.
2. New processes of production should be evaluated for safety.
3. Evaluations should have broad participation.
4. Evaluation can result in recommendations for animal testing.
5. Evaluation committee should have de facto authority over national policies on GM foods.
6. International organizations should harmonize risk (safety) assessments for both products and processes.
7. Consumer information should be scientifically based and only concerned with food safety issues.

(WHO/FAO, 1991)

These recommendations were crucial in forming a baseline approach for assessing and regulating GM crops. The first recommendation supported the product basis for regulatory oversight, while the second recommendation suggested specific oversight in the instance of novelty. The third recommendation supported the need to include 'social dimensions' in regulatory development. The fourth recommendation encouraged more pharmaceutical-type assessment procedures for approval of novel foods, while both the fifth and sixth recommendations called for the development of a harmonized, international regulatory framework to GM crops, overriding national regulations. The seventh recommendation only supported consumer information, such as labelling strategies, in instances of hazard or food-safety concern, such as the possible presence of allergens, and not for the consumers' right to know. Also in 1990, the Joint Expert Committee on Food Additives developed risk-assessment guidelines for the use of GM-material additives in foodstuffs.

The Codex Medium Term Programme of Work for the period 1993 to 1998 included the development of Codex guidelines for the evaluation of foods produced from biotechnology and the application of risk-assessment principles to the Codex work on agricultural biotechnology (Codex Secretariat, 1993). Unsurprisingly, similarly to the 1991 recommendations above, the guidelines of the Codex Secretariat supported the scientific-rationality approach to novelty-based regulations.

In 1996, a Joint FAO/WHO Consultation on Biotechnology and Food Safety was held. The purpose of this consultation was to examine the broad implications of biotechnology for food safety and to propose Codex procedures for dealing with biotechnological applications.

At the 23rd CAC Session in Rome, 28 June–3 July 1999, the USA proposed a 'Biotechnology Code' that clarified the dominant role of science in the food standards associated with products of biotechnology. This code, congruent with the 1993 Codex Secretariat guidelines, called for novelty-based regulations. Although unsuccessful, the CAC agreed instead to establish an *Ad hoc* Intergovernmental Task Force on Foods Derived from Biotechnology to examine the issues facing Codex efforts to develop biotechnology standards, codes of practice, guidelines and recommendations. The Task Force is wider in remit than the Biotechnology Code, as it will allow for broader non-science issues to be considered, such as consumer- and environmental-protection issues. Specifically, the Task Force will:

1. Elaborate standards, guidelines and other principles, as appropriate, for food derived from biotechnology;
2. Coordinate and closely collaborate, as necessary, with appropriate Codex Committees within their mandates as related to foods derived from biotechnology; and
3. Take full account of existing work carried out by national authorities, FAO, WHO, other international organizations and other relevant international fora.

The Task Force was initially given a 4-year mandate from July 1999 to July 2003 requiring a preliminary report to the 24th CAC in 2001, a mid-term report to the CEC in 2002 and a Final Report to the 25th CAC in 2003 (*AgraFood Biotech*, 1999a). The first meeting of the Task Force was held in Chiba, Japan, 14–17 March 2000, and the issues covered included: establishing the scope and priorities of the Task Force, clarifying key concepts and definitions for core principles, such as risk analysis, and examining national and regional experiences with the regulatory problems created by foods derived from biotechnology.

While the Codex appears to be committed to a product-based, novelty-focused approach to GM-crop regulations, an institutional shift may be on the way. The Codex Medium Term Programme of Work 1998–2002 contained a proposal to establish Codex measures on the application of biotechnology. It requires 'consideration of a general standard for foods derived from biotechnology or traits introduced into foods by biotechnology'. Essentially, this means that a Codex horizontal General Subject Committee on biotechnology would have to be established, essentially shifting biotechnology from its current novelty focus to a process- or technology-based focus.

In summary, the Codex employs a scientific-rationality approach to the development and integration of food-safety regulations – an approach supported by economic interests. Accordingly, the development of Codex standards for the use of biotechnology in food production has tended to reflect this approach. Yet there is significant dissatisfaction with this approach and debates have emerged over whether or not the Codex can effectively hold the line on its traditional scientific approach. In this sense, the development of international biotechnology standards at Codex remains unstable and uncertain.

D. International Plant Protection Convention

While the Codex deals with sanitary (food) standards and regulations, the IPPC deals with phytosanitary (plant) standards and regulations according to a scientific-rationality approach. The IPPC is important for the international market access of GM crops because, according to the SPS Agreement, it is the only international institution that establishes the legitimate use of phytosanitary standards as trade barriers. Its scientific-rationality approach is due to the fact that international scientific experts on phytosanitary hazards have dominated the development of international plant-health regulations. This work has been carried out far from public view. However, similar to the Codex, this is controversial, because this approach contravenes the social-rationality approach.

The IPPC is a multilateral environmental agreement – signed in 1951 and coming into force in 1952 – with a broad biodiversity-protection remit. The scope of the convention is the protection of natural flora, cultivated plants and plant products. Similarly to the Codex, the motivation for the IPPC was to develop international standards, applicable to all countries, for the protection of plant health and hence remove the fragmented collection of standards in the various jurisdictions. To achieve this, the IPPC seeks to harmonize international measures designed to prevent the introduction and spread of diseases and pests to plants and plant products. Clearly, as the environmental-biodiversity concerns about GM crops mostly include speculation on the risks, extent and consequences of vector-mediated, horizontal gene transfer between GM crops, non-GM or conventional crops, wild relatives and other natural flora and organisms, the scope of the IPPC is well positioned to focus precisely on these concerns.

The IPPC is administered through the IPPC Secretariat which is part of the FAO's Plant Protection Service. Currently, 107 governments are contracting parties to the IPPC. The Secretariat, in collaboration with both regional and national plant-protection organizations (RPPOs and NPPOs), provides a forum for the international cooperation and harmonization and the technical exchange of plant-protection information.

The IPPC plays a vital role in international trade, as it is recognized in the SPS Agreement as the institution responsible for developing international standards for phytosanitary measures affecting trade in plants and plant products. Where the SPS Agreement is concerned, the IPPC standards are considered in the limited capacity of protecting the plant health of the domestic food supply from pests and disease. In this sense, IPPC standards may be used legitimately under the SPS Agreement to restrict imports of certain plants and products produced from plants, because of the risk of transmission of diseases or pests to plants used in the domestic food supply, but not for protecting overall plant biodiversity. Therefore, while the IPPC's phytosanitary standards are crucial to the SPS Agreement from a more limited food-safety perspective, they are really about environmental protection and thus the IPPC straddles the divide between food-safety and environmental-protection measures. Consequently, the IPPC will be discussed first in its food-safety context under the SPS Agreement, followed by a discussion of its environmental context in the next section on Environmental Protection and Trade.

To accommodate the formal link with the SPS Agreement, signatories to the IPPC agreed on amendments in 1997 in order to clarify the standard-setting procedures. The 1997 amendments, captured in the New Revised Text of the IPPC, include provisions that:

1. Formalize the role of the IPPC Secretariat;
2. Update the standards-setting procedures;
3. Emphasize cooperation and the exchange of information toward the objective of global harmonization; and
4. Establish the Commission on Phytosanitary Measures (CPM). The CPM will serve as the global agreement's new governing body.

The members of the Commission are the contracting parties to the Convention. The creation of the Commission on Phytosanitary Measures (CPM), in effect, institutionalized the global role of the IPPC. The Commission meets annually to establish priorities for standard setting and the harmonization of phytosanitary measures in coordination with the IPPC Secretariat. Special sessions of the Commission may also be called. The functions of the Commission are to review the state of plant protection in the world, provide direction for the work programme of the IPPC Secretariat and approve proposed International Standards for Phytosanitary Measures (ISPMs). In this sense, the IPPC must be considered an active player in the development and integration of GM-crop regulations.

With respect to the international regulation of GM crops, the IPPC signatories develop ISPMs. The ISPMs do not explicitly address agricultural biotechnology as a process. Instead, products of agricultural biotechnology that pose a risk to plant health are within their scope. In this sense, the risk-assessment efforts of the IPPC have adopted a product-based approach to GM crops, similar to that of the Codex and congruent with the traditional trade principle of 'like' products or 'substantial equivalence'. There are currently nine ISPMs accepted by the contracting parties to the convention that reflect procedural standards.[7]

The development of ISPMs follows a three-stage procedure. First, suggestions for drafting an ISPM are made by either the Secretariat, an NPPO, an RPPO, industry participants or individuals. Draft standards are then developed and submitted to the Secretariat by the NPPOs or RPPOs. The draft standards are reviewed by the Committee of Experts on Phytosanitary Measures (CEPM), a group of phytosanitary experts from around the world that meets annually to review and comment on the suitability of documents prepared by the Secretariat. Alternatively, the draft standard may be reviewed by an international working group formed by the Secretariat. Recommendations to either develop an ISPM from the draft proposal or modify the draft proposal for further review are made. After the draft ISPM is developed, contracting parties and RPPOs are consulted. Comments are submitted to the CEPM and the Secretariat and a redrafted standard is developed. This standard is then submitted to the CPM for approval and adoption as an ISPM. The standard is published and distributed by the FAO.

The Convention is a legally binding agreement in terms of its principles of plant protection, and member governments or contracting parties are expected to adhere to the ISPMs, but standards developed and adopted are not legally binding. Similarly to Codex standards, the importance of IPPC standards is that measures that are based on the international standards do not require supporting justification. Measures that deviate from international standards or measures that exist in the absence of international standards must be based on scientific principles and evidence in order to be considered as scientifically justified under the SPS Agreement. Emergency (or provisional) measures may be taken without such analyses, but must be reviewed for their scientific justification and modified accordingly in order to remain legitimate.

[7]The ISPMs are: Principles of Plant Quarantine as Related to International Trade; Guidelines for Pest Risk Analysis; Code of Conduct for the Import and Release of Exotic Biological Control Agents; Requirements for the Establishment of Pest-Free Areas; Glossary of Phytosanitary Terms; Guidelines for Surveillance; Export Certification Systems; Determination of Pest Status in an Area; and Guidelines for Pest Eradication Programmes.

In respect of the risk-analysis framework, the IPPC supports scientific-rationality approaches, where contracting parties are obligated to be scientific in their approach to phytosanitary measures, although the initiatives include risk-assessment, risk-management and risk-communication activities. In terms of scientific risk assessment, the IPPC obliges signatories to carry out pest surveillance and monitoring and to conduct risk assessments on the introduction and spread of pests in order to provide scientific support for phytosanitary measures. The phytosanitary standards developed by the IPPC aim to establish a rules-based approach to scientific risk assessment according to the principle of substantial equivalence. In terms of risk management, many of the IPPC Secretariat's activities are consistent with the overall objectives of the FAO, including: capacity building and strengthening plant-protection infrastructures; dispute avoidance and resolution; updating and aligning plant-protection legislation; and the provision of emergency assistance programmes. In terms of risk communication, signatories must appoint a contact point to provide communication, information sharing and transparency between contracting parties and the IPPC Secretariat.

As mentioned above, the IPPC can perform an important risk-management function – a function that the Codex does not have – namely, dispute resolution. The IPPC has built-in provisions for dispute avoidance and dispute settlement in the event that measures are challenged as unjustified barriers to trade. The Secretariat provides guidance, support and information for contracting parties concerning phytosanitary measures and it facilitates the exchange of information between the parties in respect of regulatory requirements and pest status. The dispute-settlement process under the Revised IPPC provides a neutral forum for a technical dialogue on the dispute. Countries first consult bilaterally with the aim of resolving the problem. The IPPC Secretariat provides technical support and facilitates the exchange of views and information in this process. If further action is deemed necessary, the disputing parties can request that the Director-General of the FAO form an expert panel to review the situation and recommend a course of action. Although the dispute-settlement process in the IPPC is non-binding, the results of the process can be expected to have substantial influence in disputes that may be raised at the WTO under the SPS Agreement. This is because the IPPC Secretariat both nominates experts for WTO dispute panels and provides technical background for the panel. Once a dispute is brought to WTO, the decision will be legally binding and can have serious economic and political consequences. Therefore, the IPPC encourages governments to begin with consultation and a technical exchange, with the aim of dispute avoidance at this technical stage before the political stakes are raised in a trade dispute.

The effectiveness of the dispute-avoidance and dispute-resolution provisions of the IPPC remains unknown. In the event that trade barriers to GM crops are supported by phytosanitary measures, then it is possible that the dispute provisions in the IPPC are used to resolve the social regulatory barriers, rather than the WTO dispute-settlement mechanism. However, the problem would be that the IPPC is thrust fully into the realm of risk management and top-down regulatory integration, rather than remaining an international forum for establishing science-based principles for risk assessment and for providing risk information. Dispute-settlement decisions on appropriate domestic regulations will necessarily create winners and losers. This, in turn, creates dissatisfaction among the losers towards an international institution that is supposed to be developing consensual, coordinated phytosanitary measures but is instead disciplining domestic measures according to the scientific-rationality approach.

In short, it is easy to understand why economic interests support the IPPC as the international organization responsible for developing internationally harmonized phytosanitary standards for GM crops. Its remit to promote the health of natural flora, cultivated plants and plant products allows it considerable scope to cover the various environmental-biodiversity concerns that have been levelled against GM crops. Yet its adherence to scientific rationality limits consideration of 'other legitimate factors'. Further, its built-in dispute-settlement provisions would focus on science-based evidence of risk and it would avoid having to settle phytosanitary disputes at the WTO. However, the criticisms of the science basis of the Codex all apply equally to the IPPC, leaving its role in an international regulatory regime for GM crops uncertain.

To summarize the approach of the international trade regime for dealing with food-safety regulations, a two-prong strategy is consistently used. Safety-related measures (i.e. hazard or risk) are separated from non-safety measures (i.e. quality or packaging). In the case of the former, countries may take unilateral actions to prevent food-safety risks from entering their territory, subject to scientific evidence of risk. In the case of non-safety measures, they are subject to the traditional principle of non-discrimination, and violations would be in contravention of the international trade regime.

3.2.2 Environmental protection and trade

To date, much of the discussion on GM crops and trade has focused on the relationship between food-safety measures and the rights and obligations of trade agreements (Isaac and Woolcock, 1999b; Perdikis *et al.*, 1999). Yet environmental-biodiversity concerns are a big part of current consumer concerns about GM crops, and they create pressures to impose environmental-protection measures, which become social regulatory barriers. The trade concern is that the use of environmental-protection measures is less disciplined under international agreements than that of food-safety measures. Hence, it is important to understand how environmental-protection measures are dealt with at the WTO. It will be argued that, in general, economic interests underlying the international trade regime support a rules-based and science-based approach to the development and integration of environmental regulations.

Environmental-protection rules have been less rules-based and science-based under international agreements. Perhaps the most important reason for this situation is that the science of food toxicology is more advanced than the science of predictive ecology, essentially limiting the extent to which environmental regulatory development could rely upon science. However, with the Uruguay Round Agreement, economic interests have, in fact, begun to formalize environmental-protection measures under science-based trade rules, in a manner similar to the formalization of food-safety rules. This has come about in two distinct ways. First, there has been an attempt to clarify safety-related environmental measures under the SPS Agreement's link to the IPPC and to clarify non-safety-related environmental measures under the TBT Agreement. Secondly, there has been an attempt to demarcate the line between trade agreements and multilateral environmental agreements.

A. Processing and production methods

The traditional focus of GATT 1948 was on products, not PPMs, because products, not processes, are traded and cross international borders. Increasingly, however, it became clear that the manner in which a product was processed or produced was relevant to domestic regulations and hence was relevant for the access of foreign products to domestic markets. While recognizing the importance of PPMs in domestic environmental regulations, the international trade regime sought to clarify their legitimate use according to the two-pronged approach used in dealing with food-safety regulations (Table 3.3). Safety-related PPM-based environmental measures are addressed under the SPS Agreement through the IPPC, similarly to the way food-safety measures are addressed under the SPS Agreement through the Codex. Non-safety-related PPM-based measures are generally addressed under the TBT Agreement, although some clarification is required. The key point is that, similarly to food-trade regulations, the aim has been to disentangle safety measures from non-safety measures and make the former science-based and the latter subject to the traditional trade principle of non-discrimination.

Table 3.3. Categories of environmental PPMs.

Safety-related PPMs	Non-safety-related PPMs
SPS Agreement: Compliance requires: 1. IPPC standards on phytosanitary measures 2. Scientific evidence of risk to deviate from the non-discrimination principles according to the IPPC approach to the risk-analysis framework	TBT Agreement: Compliance requires that the environmental measures meet the principle of non-discrimination

Under the SPS Agreement, the use of safety-related phytosanitary PPM-based environmental measures are subject to either the phytosanitary standards developed by the IPPC or the IPPC risk-analysis framework to standard-setting – that is, the phytosanitary standards and standard-setting procedures of the IPPC must be considered as the parameters for the permissible phytosanitary measures that WTO members can legitimately adopt. This creates a rules-based and science-based approach to safety-related environmental-protection regulation, similar to the Codex approach to food-safety regulation.

Non-safety-related PPM-based environmental measures are not subject to the same degree of scientific justification as safety-related measures. Without this scientific basis, non-safety-related measures are disciplined under the TBT Agreement, according to the traditional trade principle of non-discrimination. With respect to environmental measures pertaining to GM crops, the main TBT issue is the permissible use of non-safety trade-restriction measures, based on the PPMs.

The relationship between non-safety-related PPMs and international trade was investigated in both the Tokyo and the Uruguay Rounds of multilateral trade negotiations. In the Tokyo Round, Contracting Parties differed as to how they thought PPMs should be dealt with. On the one hand, some Contracting Parties, including major agricultural exporters, sought to have PPMs included in the TBT Code to prevent circumvention of the GATT trade principles by technical regulations. The aim was not to prevent the use of PPMs in product standards, such as labelling requirements, but to have some discipline on their use. Other Contracting Parties, such as developing countries, did not want to have PPMs included in the Code at all, so that any PPMs used to ban trade would be in violation of the TBT Code. Most developing countries did not want any linking of trade and the environment, because often PPMs are considered to be a source of comparative advantage. They point out, with some justification, that the industrialized countries developed without environmental controls.

Now that they have all the benefits conferred by industrialization, especially higher incomes, they can afford to demand more income-elastic environmental protection through PPM-based trade measures. This makes the industrialization process harder for the South. Such measures may favour particular PPMs or technologies that are not employed by and/or not available to developing-country producers. Also, products crucial to developing countries (i.e. paper products/timber) are disproportionately covered by PPM measures. In other words, if PPMs favour new technologies, then the capacity gap means that the less technologically advanced will continue to lag behind.

In order to deal with the differing views on the legitimate use of PPMs in the trading system, an important compromise was struck. First, PPMs were divided into two types; product-related PPMs and non-product-related PPMs (OECD, 1997c). This was done in order to reclaim some of the traditional division between products and processes under the GATT. Secondly, the use of product-related PPMs was subjected to the dispute-settlement provisions of the TBT Code (TBT Code, Article 14:25).

In the Uruguay Round, the USA, along with other agricultural exporters, again tried to have product- and non-product-related PPMs disciplined, this time under the binding TBT Agreement. Yet, at the end of the Round, the only change was that product-related PPMs would be covered by the TBT Agreement, not non-product-related PPMs. This means that the rights and obligations of the TBT Agreement only apply to those PPM measures that have an effect on the safety or quality attributes of the final product. Accordingly, there are no permissible circumstances for employing non-product-related PPMs. As will be shown below (Chapter 5), GATT and WTO dispute decisions have consistently concluded that trade restrictions based on the non-product-related PPMs of traded goods violate the trade principles of non-discrimination and that they result in an unjustifiable extraterritorial extension of one member's environmental preferences upon another member (Table 3.4).

In respect of food trade, non-safety product-related PPMs can include, for instance, the type of veterinary practices and quality-assurance systems that may be employed in a beef production system, because these PPMs may affect the safety and quality of the final beef products. That is, product-related PPMs are associated with the consumption or use stage of the product and may cause negative consumption externalities. According to the TBT Agreement, coverage includes 'product characteristics or their related processes and production methods', but only in so far as they involve 'consumption externalities' (TBT Agreement, Annex 1). Therefore, only consumption externalities associated with product-related PPMs are within the scope of the TBT Agreement. Further, and similarly to the SPS

Table 3.4. Further categories of environmental PPMs.

	Non-safety-related PPMs	
Safety-related PPMs	Product-related	Non-product-related
SPS Agreement: Compliance requires: 1. IPPC standards on phytosanitary measures 2. Scientific evidence of risk to deviate from the non-discrimination principles according to the IPPC approach to the risk-analysis framework	TBT Agreement: within scope Compliance requires that the environmental measures meet the principle of non-discrimination	TBT Agreement: outside scope No legitimate use according to international trade rules

Agreement, there are limits to the permissible use of product-related PPM-based measures that members may enact. Under the TBT Agreement, the legitimate deviations are not as precise as those in the SPS Agreement and can include ‘different social objectives and priorities attached to environmental protection’ (OECD, 1997c).

On the other hand, non-safety, non-product-related PPMs in the trade of agricultural crops are generally associated with the agronomic system, but have no influence on the end-product. For instance, the technologies employed or the soil cultivation or conservation strategies used are non-product-related PPMs. Although not responsible for consumption externalities, these PPMs may cause negative-production environmental externalities. Yet these PPMs are outside the scope of the TBT Agreement and are the sovereign domain of the member government. From a trade perspective, such measures must follow the principles of non-discrimination.

From an environmental-protection perspective, the problem with the fact that non-safety, non-product-related PPMs are outside the scope of the Agreement is that the pursuit of sustainable development and the protection of biodiversity often focus on production externalities and result in pressures on domestic governments to establish such measures pertaining to non-product-related PPMs, which can become regulatory market-access barriers. Specifically, GM crops are associated with concerns about their impact upon the environment and this concern has led to opposition in member states to the domestic environmental release of GM crops. It is reasonable to suppose, however, that this may even lead to a member enacting trade measures based on the non-product-related PPMs of GM crops grown elsewhere and imported into the member country.

Despite the importance of the issue, the compatibility of PPM-based environmental measures with the rights and obligations of the TBT Agreement remains uncertain (Caldwell, 1998). The fragmentation of PPMs into various categories based on vague criteria makes it difficult to understand which environmental regulations are in compliance with international trade obligations and which are in contravention.

Moreover, in respect of GM crops, an additional unresolved issue exacerbates the trade uncertainty. It can be expected that this issue will be an important focus of the TBT Committee, as trade issues associated with GM crops arise. The unresolved issue is whether or not GM crops can even be categorized as having different PPMs from conventional, non-GM crops. Consider that genetic modifications are made to the cells of the crop, which are then cultured into seeds. In the case of production-trait varieties, the GM crops are then grown in exactly the same agronomic system as the conventional varieties. Then the crops are harvested and sold into the same processing and distribution system as non-GM varieties. Accordingly, GM crops that do not require new or different agronomic or processing systems would not be grown under process or production methods different from those for conventional crops and it appears that there would be no difference between the non-product-related PPMs of GM crops and non-GM crops. Furthermore, as many of the approved GM varieties to date have been approved as like products or as substantially equivalent to conventional varieties, there are no differences in product-related PPMs either and there are no consumption externalities to consider. In this case, the TBT Committee will have to decide, first, on the applicability of PPMs to GM crops and, secondly, on the legitimacy of the principle of substantial equivalence.

Of course, the TBT Committee does not have to make this decision proactively. Instead, it could allow differences of opinions among members to escalate from social regulatory barriers to trade tensions, followed by trade disputes brought to the WTO's dispute-settlement body. In Chapter 5, several WTO disputes based on food-safety and environmental-protection measures are examined, in order to extrapolate how the dispute-settlement body may deal with such GM-crop controversies.

B. Trade and multilateral environmental agreements

Another crucial issue for environmental protection and trade deals with the often difficult relationship between the international trade regime and multilateral environmental agreements. In November 1999, the WTO's Committee on Trade and the Environment (CTE) published a report on the environmental effects of trade (WTO, 1999a). The report argues that, although trade, *per se*, is not harmful to the environment, it is not beneficial in all cases. The report concurs with environmental concerns on four issues. First, subsidies for export-orientated commercial activities, such as farming and fisheries, and subsidies for fossil-fuel-driven activities create incentives for economic activities that are not congruent with sustainable development. Secondly, voluntary third-party eco-labels are a valuable tool for providing consumer information about the use of acceptable non-safety, non-product-related PPMs. Thirdly, the CTE supports a comprehensive review by members on the environmental effects of trade in order to increase the transparency of environmental impacts. Fourthly, the CTE recognizes that the WTO, although accountable to member governments, is not accountable to or transparent or accessible for non-members, such as environmental organizations. The report suggests that the WTO must address these four areas if it is to address the harsh criticisms levelled against it by environmental organizations.

The report argues, however, that the primary goal of the WTO is to enhance trade liberalization and that the WTO is not the proper venue for environmental concerns. For instance, first-best environmental policies, not second-best trade policies, must tackle negative environmental effects at source. Consequently, the report concludes that trade rules, in pursuit of free trade, cannot be altered to accommodate environmental objectives. In this sense, there is only so much that the WTO is able and willing to do to meet environmental objectives. These limitations often give rise to negotiations for multilateral environmental agreements (MEAs) outside the WTO framework.

MEAs are intergovernmental agreements on environmental objectives that reflect a regulatory coordination strategy of social regulatory integration. It is crucial to understand their relationship with the traditional trade approach, because generally the social-rationality approach prefers the deeper social integration of the MEAs over the shallow, economic integration of trade agreements.

MEAs or the smaller regional environmental agreements (REAs) generally focus on the non-product-related PPMs that are outside the TBT Agreement. There are about 180 and they include the Kyoto Protocol, the Basle Convention on the Transboundary Movement of Hazardous and Toxic Waste and the Montreal Protocol. Generally, MEAs are agreements between like-minded countries. It has been argued that MEAs require a shared view of environmental harm that is often based on 'moral values, ethical or cultural preferences or environmental choices which lack a scientific basis' (OECD, 1997c). Also, as the demand for environmental-protection measures is income-elastic, such measures tend to be supported by higher-income, developed countries, while lower-income, less developed countries generally view them as a threat. As mentioned, the reason for this is comparative advantage, where many developing countries view their PPMs as a source of advantage and resist the idea of limiting this advantage through MEAs (OECD, 1993).

MEAs aim to focus domestic environmental policies on agreed global objectives, such as ozone depletion. MEAs tend to come into conflict with trade only when non-signatories are affected by the obligations that have been negotiated by signatories. For instance, a trade dispute requires a complaint to the WTO. Members can have and keep policies that contravene the rights and obligations of WTO agreements, provided no other member complains. Signatories to an MEA would all share the same environmental-protection philosophy, so that, even if domestic environmental regulations in one signatory are trade-restricting, other signatories would not complain because they appreciate the broader environmental objective. A non-signatory, however, who does not share the same environmental-protection philosophy but is affected by the trade-restricting measure is more likely to complain about the measure to the WTO.

Yet the problem with MEAs is not always between signatories and non-signatories. The Cartagena Protocol on Biosafety (discussed in Chapter 4) is an example of the difficulties that arise when MEAs encroach upon international trade. During negotiations, the Cartagena Protocol threatened to have a significant impact upon the trade of GM crops (Isaac and Phillips, 1999c). As a result, the Miami Group of agricultural exporting signatories effectively blocked agreement of the protocol in February 1999, indicating the dominance of trade concerns over environmental concerns.

In terms of the regulatory integration strategy, MEAs consistently follow a regulatory-coordination approach based on consensual negotiations involving wide participation, which is a time-intensive process. It is unlikely that an MEA can be negotiated rapidly enough to deal with the environment and trade concerns arising from PPM-based environmental measures in order to curtail the pressure for unilateral action and a trade complaint to the WTO.

As mentioned, the CTE has investigated the relationship between MEAs and the rights and obligations of the trade agreements. Specifically, it has examined the permissibility of non-product-related PPM-based measures that differentiate between physically or functionally similar products that exhibit no consumption externalities. Currently, no judgement has been made.

3.3 Conclusions

In this chapter, the regulatory development and regulatory integration approaches for GM crops supported by economic interests were examined. It was revealed that at the heart of the support is the demand for certain and predictable rules for the development, commercialization and international market access of GM crops. Congruent with the economic perspective and the scientific-rationality approach, economic interests support the correction of real market failures permitting economic growth and development, which in turn lead to higher income-elastic social regulations.

Economic interests support regulations that ensure technological progress through capacity building. Capacity permits the development and commercialization of GM crops, securing both private and public economic gains. Economic efficiencies are gained through the vertical and horizontal integration of the GM-crop applications.

The commercialization of GM crops requires stable and predictable regulatory rules, and economic interests support a science-based framework for these rules. This includes rules for product development, testing, field trials and market approval, as well as rules for intellectual property protection.

In addition, as commercialization at the international level is required to recoup R&D costs, economic interests also support international market-access rules that are congruent with a science-based framework. They support the WTO approach to clarifying the trade-compliant use of food-safety and environmental-protection measures through the SPS Agreement and its links with the Codex and the IPPC. The key feature is that the traditional trade approach attempts to disentangle safety-related measures from non-safety-related measures. In order to be considered legitimate, safety-related food- and environmental-protection measures must be scientifically justifiable, while non-safety-related measures must meet the trade principle of non-discrimination.

The traditional trade approach, as outlined in the SPS and the TBT Agreements, along with the accompanying international institutions, such as the Codex and the IPPC, provides a coherent approach to international regulatory development and integration. Member countries may unilaterally impose social regulatory barriers to GM crops, but only with appropriate scientific justifications.

Social Interests 4

In the previous chapter, the regulatory development and integration approaches supported by economic interests were examined. It was argued that these interests have dominated the development of the international trade regime with the result that it reflects a scientific-rationality, rules-based framework for regulatory development and encourages a regulatory-competition strategy of integration among members. Yet, in recent years, this framework has come under considerable criticism from non-economic or social interests, who support a very different approach to regulatory development and integration.

The objective of this chapter is to examine the regulatory development and regulatory integration regimes supported by the social interests, in order to understand their criticism of the traditional trade-diplomacy approach. Broadly, this group is composed of various national non-governmental organizations (NGOs) and international non-governmental organizations (INGOs), where three types of social-interest organizations may be identified, according to the concerns they predominantly focus on: consumer, environmental and social-development organizations. This is, however, only an illustrative categorization. In most cases, it is difficult to define social-interest groups neatly as one of the three types because, regardless of their primary mandate, many of the groups include consumer, environmental and social-development advocacy.

It will be argued that these organizations generally support regulatory development and integration approaches that are socially responsive, reflecting the social-rationality approach to regulatory development and a regulatory coordination strategy for integration.

4.1 Regulatory Development and Integration

Prior to discussing each of three social-interest organizations separately, it is important to develop the categorization of these groups (Table 4.1). At the broadest level, the first distinction to draw is the difference between 'state' organizations and 'non-state' organizations. The former are organizations whose memberships are composed of nation-states and are generally referred to as international governmental organizations (IGOs) and include trade IGOs, such as the World Trade Organization (WTO), environment IGOs, such as multilateral environmental agreements (MEAs), including the Convention on Biodiversity (CBD), and social-development IGOs, such as the United Nations' World Health Organization. On the other hand, the term NGOs is often used synonymously with other terms, such as civil-society organizations (CSOs), private voluntary organizations (PVOs) or independent-sector organizations (ISOs).

According to Table 4.1, NGOs are composed of both commercial organizations and CSOs, which are then further categorized as consumer, environmental and social-development organizations. Both commercial organizations and CSOs can exist at the national or the international level. The categorizations are not free of controversy. For instance, many commercial organizations would argue that they are in fact CSOs and not a separate category. Similarly, it has been noted that some public officials have suggested that they too represent civil society (Shamsie, 2000). Other categorizations suggest that 'NGO' does not include commercial or labour organizations (Hay, 2000). Further,

Table 4.1. Categorization of social interests (from Isaac, 2001). This categorization was influenced most by the United Nations (1995) definition of non-state actors.

Non-governmental organizations (NGOs)			
	Civil-society organizations		
Commercial organizations	Consumer	Environmental	Social development
National Canadian Manufacturers Association	National Consumers' Association of Canada; National Consumers Council (UK)	National Canadian Environmental Network; Soil Association (UK)	National Canadian Labour Congress; National Rifle Association (USA)
International International Chamber of Commerce	International Consumers' International	International Friends of the Earth; Greenpeace	International Third World Network; Christian Aid

many CSOs would reject the notion that an organization such as the USA-based National Rifle Association is a CSO.

In this study, social interests are synonymous with CSOs and there are three broad categories to consider: consumer, environmental and social-development interest groups. All three tend to share an important focus on the social-rationality approach to regulatory development and integration. In addition, they all reject two important premises of the economic perspective. First, economic analysis generally assumes that technology and innovation are crucial elements of economic growth and social welfare.[1] Yet social interests argue that technology is not an inherent factor in social welfare. Instead, it is the application, management and distribution of technology that can lead to increases in social welfare or, in the case of misapplied, mismanaged and poorly distributed technology, decreases in social welfare. Secondly, the economic perspective tends to treat consumers as economic agents, where price is the only aspect of genetically modified (GM) crops that they are concerned about. For instance, if prices fall, there are consumer-welfare gains (Moschini *et al.*, 1999). Social interests argue that consumers are not just economic agents but that they have legitimate concerns beyond just economic concerns, which must be included in evaluations of technology. In the UK, these concerns have led some consumers to actually spend more money on food purchases in order to avoid products made from GM crops and, presumably, to increase their welfare (discussed more fully in Chapter 7).

Accordingly, social interests argue that regulatory development must not simply be an exercise of correcting market failure in order to maximize technological progress. Instead, it must be responsive to social preferences and concerns, regardless of the economic rationality of those concerns. In addition, regulatory integration must be sensitive to divergent social normative frameworks between jurisdictions and must avoid sacrificing social dimensions in the pursuit of economic efficiency and competitiveness, creating a so-called 'hollowing-out' of the nation-state (Picciotto, 1996). To avoid this, social interests support a regulatory-coordination approach to deeper integration.

There are five similarities between the three types of social-interest groups that are relevant to the development and integration of GM-crop regulations. First, they hold a particular belief system or frame of reference, which tends to be based on normative beliefs that are not subject to scientific rules of evidence and debate (Salamon, 1994, 1995). The belief systems of the three types of social interests will be discussed separately below, but it is important to note that some organizations can feel so strongly about their normative beliefs that they are willing

[1]Although not all economists hold this view. For an insightful discussion of the 'steady-state' or zero-growth economy, see H. Daly (1997) 'Towards a new economics: questioning growth', excerpted from Daly (1971) (http//csf.colorado.edu/authors/hanson/page41.html).

to take illegal action, as evidenced by the destruction of UK field trials of GM rape-seed (see Chapter 7). Further, the strength of the belief system may effectively limit the ability of social-interest organizations to participate in activities that demand compromise (Simmons, 1998). For instance, it has been argued that German environmental groups in a 1991 Participatory Technology Assessment of herbicide-tolerant GM crops 'had ambivalent feelings about being involved in a procedure in which they could not control the findings' (van den Daele *et al.*, 1997). In fact, van den Daele *et al.* (1997) note that these organizations withdrew from the Participatory Technology Assessment and released their version of the findings prior to the release of the final, consensus-based report. In contrast, economic interests tend to have more fluid belief systems reacting to market-demand signals, so that, when demands change, economic policies change, as evidenced by the shift to non-GM ingredients among UK supermarkets (see Chapter 7).

Secondly, for social-interest organizations, the public is the audience and these organizations tend to be very effective at conveying their position in an easily understandable manner (Clark, 1991; Kobrin, 1998). For instance, these interests have traditionally had little influence on the trade-policy process. Indeed, to increase their policy voice, they have found populist and sensational ways of circumventing traditional trade-diplomacy channels, as evidenced by the 1999 WTO Ministerial Meeting in Seattle and the 2001 Summit Negotiations of the Free Trade Area of the Americas in Quebec City. Hence, social-interest organizations are important actors in shaping public opinion, which in turn has an important influence upon regulatory development and the prospects for regulatory integration. The public orientation of these organizations is in conflict with the traditional trade-diplomacy strategy of closed-door negotiations that are neither publicly transparent nor accessible (Hay, 2000; Stairs, 2000; Wolfe, 2000). It is also in contrast to the typical scientific-rationality approach used, for instance, by the international scientific community (i.e. Codex and the International Plant Protection Convention (IPPC)), whose work is undertaken far from the public view and at a technical level not easily understood.

Thirdly, they share the objective of promoting sustainable development (International Institute for Sustainable Development, 1994). There has been a general international shift in the interpretation of 'development' as a goal of government policy (Enders, 1997). The traditional view of development is rooted in the economic perspective of continued growth and prosperity permitting higher income-elastic social regulations. Development now has taken on a decisively socioeconomic character, as it is now termed 'sustainable development', interpreted as 'development that meets the needs of the present without compromising the ability of future generations to meet their own needs' (WCED, 1987). While this term does lack precision it is understood to mean a shift in policy focus from just economic to socio-economic gains (Najam, 1996).

Fourthly, the social-interest organizations share concerns, in general, about democratic-deficit and accountability issues associated with international institutions (Johnson, 2000; Shamsie, 2000). For instance, these organizations have limited resources with which to participate in the myriad international institutions for GM-crop regulatory development (i.e. WTO, the Codex Alimentarius, the IPPC, the Organization for Economic Cooperation and Development, etc.) and they are concerned that their perspective may not be appropriately addressed when international standards are being developed. This concern was given credence at the 1999 G7/G8 Summit in Cologne, Germany, where the Transatlantic Business Dialogue (TABD) (see Chapter 5) was allowed to participate, but the Transatlantic Consumers' Dialogue (TACD) and the Transatlantic Environmental Dialogue (TAED) were not (Transatlantic Consumers' Dialogue, 1999b).

The fifth shared concern of the social-interest organizations lies with the international economic-integration approach of traditional trade diplomacy (Enders, 1997; Scholte *et al.*, 1998; Isaac and Woolcock, 1999a). They do not support the WTO as the dominant international institution for the integration of food-safety and environmental-protection regulations, because they believe that it fails to deal appropriately with the social-rationality approach. In fact, a recent petition headed by the environmental group Friends of the Earth and signed by over 1200 other social-interest groups from 85 countries was presented at the 1999 G7/G8 Summit in Cologne, Germany. The petition demanded a complete review of all the WTO's Uruguay Round commitments prior to embarking on another round of multilateral trade negotiations. The argument was that the WTO is 'built on an outdated model of swapping tariffs, when it should really be based on a competition-based system of regulation and liberalisation [sic]' (Evans, 1999). Also, the dissatisfaction of civil society with 'forced' regulatory integration through the traditional trade approach has been incisively summed up, in respect of the Appellate Body decision in the beef-hormones dispute:

> The WTO Dispute Settlement Body has, inter alia, placed the World Trade Organization in charge of determining the legitimacy of domestic health regulations; misinterpreted the provisions of the SPS text allowing countries to determine the level of appropriate risk for their citizens; favoured frequently lower international standards over higher domestic standards; dismissed the precautionary principle as a legitimate basis for health and environmental policy; and, destabilized the international trade regime by inserting itself into a dispute in which it lacks the necessary expertise and competence to adjudicate.
>
> (Caldwell, 1998)

The general rejection of the WTO's regulatory integration approach and its perceived failure to adequately deal with broader, social concerns, such as sustainable development, has resulted in a dramatic convergence of opposition to the WTO, where the loudest demand is for greater openness of the trade-policy system. While openness is conceptually a win–win situation for citizens and policy-makers, it is fraught with difficulties in its implementation, and many crucial debates must be resolved before a real increase in citizen engagement in the trade-policy advisory system is achieved (Isaac, 2001).

Despite the similarities, however, it is inaccurate to portray consumer, environmental and social-development organizations as a homogeneous group. Instead, there are important differences among them. In fact, these differences raise doubts about the compatibility of their belief systems and the potential for a lasting, cohesive opposition to GM crops. Below is an assessment of those social-interest organizations that have played very public roles in the policy debates associated with the development and integration of GM-crop regulations. This assessment is intended to be an illustrative survey and not a comprehensive catalogue. In addition, the discussion will primarily focus on those organizations that are international, but many national organizations have also established an effective presence in the regulatory-policy debates, so that the distinction between international and national social-interest organizations has become blurred.

4.1.1 Consumer organizations

The differences between the three types of social-interest organizations are especially distinct between consumer organizations and the other two. Consumer organizations advocate a comprehensive range of consumer objectives, including both economic and social concerns. Consider the four types of consumer concerns associated with GM crops identified in Chapter 2: economic, human-safety and health concerns, biodiversity concerns and moral, ethical and religious concerns. Consumer organizations have a traditional focus on the consumer's economic concerns, where they have long sought to ensure competition in markets, bringing the economic benefits of reduced price and improved quality, choice and service. Yet, while the economic concerns remain important, consumer organizations have never lost sight of the fact that consumers are not just economic agents. Consequently, they have also traditionally balanced the economic objectives with consumer health and safety objectives. The focus on sustainable development has also brought environmental-biodiversity concerns fully into the agenda of consumer organizations. Finally, social norms, including moral, ethical and even religious concerns, have played an important role in establishing the policy parameters within which the economic objectives of consumer organizations are pursued. Therefore, consumer organizations have first-hand knowledge of the difficulty associated with balancing economic and social interests across the broad range of concerns. In fact, the comprehensive range of objectives is precisely why consumer organizations in general support freer trade because of the economic benefits of competition. Yet increasingly this support is qualified; freer trade must not come at the expense of lower standards for food safety and environmental protection. The other two types of organizations, on the other hand, are more singular in purpose and do not face the same kind of balancing act.

Consumer organizations essentially have two primary concerns associated with the regulation of GM crops: first, to ensure that regulatory development is fully focused on the consumers' right to know; and, secondly, to ensure democratic accountability and transparency of international regulatory integration strategies, focused on deeper, social integration. Among these organizations, neither trade nor GM crops are bad *per se* and, in fact, most consumer organizations are not calling for a complete moratorium on GM crops. Instead, they are concerned that the 'economic' perspective in favour of technological progress, national competitiveness issues and regulatory competition is at the expense of stringent consumer and environmental regulations. They argue that, when consumers have concerns about a product like GM crops, regulatory development and integration must be socially responsive enough to react to these concerns, regardless of the economic rationality of such a regulatory reaction. Moreover, they argue that international institutions charged with developing international standards, such as Codex, cannot just be accountable to national governments, because national governments, lobbied by industry and in pursuit of competitiveness, may collectively ignore broader consumer concerns. Nor can they be only science-based, in the face of a large information gap and credence concerns. Essentially, the consumers' right to know about perceived risks must be the primary focus, regardless of the lack of scientific justification for such a perception.

Accordingly, consumer organizations tend to share the same positions on trade and GM crops in both North America and the European Union (EU), despite the asymmetrical consumer acceptance identified in Chapter 2. In the EU, the European consumers' organization, Bureau Européen des Unions de Consommateurs (BEUC), argues that the primary goal of the EU's genetically modified organism (GMO) regulations (90/220 Directive) should be to satisfy consumer concerns and 'thus, encourage consumer acceptance' (BEUC, 1999). This includes demands for greater consumer consultation at all stages of the GM-crop approval process. The UK-based National Consumers Council (NCC) argues that Codex must be committed to a full freedom-of-information policy and greater consumer participation in expert committees (National Consumers Council, 1998). Similar demands can be found in North America. The USA-based Consumers' Choice Council – an association of USA-based NGOs across all three types: consumer, environmental and social-development organizations – argues that the consumers' right to know must be the dominant justification for labelling schemes (Consumers' Choice Council, 1999). The Consumers' Association of Canada argues that consumer concerns, however broad, should dominate regulatory development through wide participation (Consumers' Association of Canada, 1994, 1999). More recently, the TACD has argued for assurances that trade liberalization will not compromise high food-safety and environmental-protection standards and that negotiations will encompass the broader concerns of consumers, rather than be focused on traditional trade interests (Transatlantic Consumers Dialogue, 1998). In respect of GM crops, the TACD has argued for a mandatory labelling scheme, based on the consumers' right to know, in Codex standard-making procedures (Transatlantic Consumers Dialogue, 1999a).

Thus, consumer organizations tend to call for greater democratic participation and accountability of regulatory and trade-policy development within international institutions. This includes greater participation of all stakeholders, not just scientists and national delegations, in the formation of international standards and in trade-dispute decisions. They also call for consumer- and environmental-safety regulations that keep pace with technological innovation, while remaining focused on the consumers' right to know (Consumers' Association of Canada, 1994; BEUC, 1999; Consumers' Choice Council, 1999; Evans, 1999).

Notable consumer organizations involved in issues of trade and GM crops include: Consumers' International (http://www.consumersinternational.org/); International Association of Consumer Food Organizations (http://www.cspinet.org/reports/codex/iacfosum.html); BEUC (http://www.beuc.org/); Consumers' Choice Council, USA (http://www.consumerscouncil.org/); Consumers' Association, UK (http://www.which.net/); and the Consumers' Association of Canada (http://www.consumer.ca/).

4.1.2 Environmental organizations

Environmental organizations differ from consumer organizations because, in respect of the four types of consumer concerns, they target almost exclusively the environmental biodiversity concerns associated with GM crops. Unavoidably, of course, both human-safety concerns and social norms also influence the position of environmental organizations. Yet what is interesting is that this position often disregards the legitimate and crucial economic factors. The benefit of having a virtually singular objective is that positions are not balanced between competing objectives. Instead, environmental protection is more of a belief system based on sustainable development, which tends to be incongruent with compromise or concession.

There are two general aspects of the environmental belief system that are relevant for the international trade of GM crops. The first is that, according to the belief system, trade liberalization and sustainable-development objectives are not compatible. There have been many calls by environmental organizations for a revision of the WTO to focus less on trade liberalization and more on sustainable development. In the event that the two objectives clash, they hold that environmental principles should dominate trade principles. Without such a shift, it is argued that the WTO should be dissolved, or at least ignored.

The second is that intensive agricultural production is incompatible with sustainable-development objectives. Most environmental organizations support, instead, an organic system of production. Hence, GM crops targeted to the intensive agronomic system contravene the belief system in principle, regardless of the attributes of these crops. The result has been calls for complete bans or moratoriums on the technology, because it has been commercialized for the intensive agricultural system. Environmental organizations in the UK have suggested that the government must decide between either organic production or GM crops in an intensive system. Yet, from a regulatory policy perspective, this belief is problematic, because there has been insufficient justification as to why GM crops and organic production are mutually exclusive. The real choice is between organic production and intensive production, not between organic and GM crops. GM crops have simply become the vector of concern because of their application as products of multinational corporations designed for the intensive system.

From the environmental perspective, it is argued that GM crops have been developed by multinational corporations, which play by the trade rules of the WTO (*Ecologist*, 1998). Yet underlying this criticism is a crucial policy issue. Clearly, the concern lies not with the technology *per se* but with the way it has been applied and, more importantly, who has applied it. Moreover, the perspective is not really anti-globalization either, as many wish to see the development of a strong, rules-based international institution, such as the Biosafety Protocol (BSP), with a global sustainable-development remit that dominates trade-liberalization efforts, responsible for establishing socially responsive regulations that employ the social-rationality approach.

Notable environmental organizations in opposition to GM crops include: International Institute for Sustainable Development (http://iisd.ca/); Friends of the Earth (http://www.foei.org/); Greenpeace (http://www.greenpeace.org/); World Wide Fund for Nature (http://www.panda.org/); Sierra Club (http://www.sierraclub.org/); European Environmental Bureau (http://www.eeb.org/); GenetiX Snowball UK (http://www.gn.apc.org/pmhp/gs/); Research Foundation for Science and Technology, India (http://www.vshiva.net/Nine.htm); and the Grass Roots Environmental Effectiveness Network (http://www.defenders.org/grnhome.html).

4.1.3 Social-development organizations

Social-development organizations, as the name suggests, focus on social development at the national, regional and/or international level, according to a particular normative belief system (Salamon, 1995; Najam, 1996). In this sense, they begin from the fourth type of consumer concerns (moral, ethical and religious concerns) and this foundation then shapes the social-development policies. Similarly to environmental organizations, social-development organizations are dissatisfied with the international trade of GM crops, although the reasons differ. In respect of trade and the WTO, they believe that the rules for international trade are designed to benefit the industrially powerful at the expense of the industrially weak and they argue that the WTO has become a vehicle for multinational corporations to dominate global production. In other words, they would support trade liberalization if it were used explicitly to pry open the affluent markets of developed countries so that producers in less developed countries could earn foreign currency.

Specifically, development, commercialization and social-development organizations argue that the international trade of GM crops furthers the divide between the north and the south. The socio-economic implications of agricultural biotechnology for development are explicitly linked with capacity, and developing countries have much at stake. Agriculture is often the largest sector in terms of income, employment and foreign-exchange earnings. While some argue that GM crops will be beneficial for the economic prospects of developing countries, others argue that they will have serious adverse impacts upon development.

Some supporters of GM crops have promoted a global-welfare promise of GM crops. They argue that agricultural biotechnology will greatly enhance agricultural productivity in less developed countries. Indigenous seed varieties can be specifically modified to the ecological conditions that currently limit production, resulting in greater yields of better quality, with a reduced need for chemical treatments. An economic study of the adoption of GM virus-resistant potatoes in Mexico concludes that GM crops can be conducive to and beneficial for small-scale farming operations in less developed countries (Qaim, 1999). From an economic-development perspective, agricultural productivity may be increased without the environmental degradation of land conversion and chemical treatments. Environmentally sustainable agricultural production can meet the food demands of a rapidly increasing global population. The argument follows, then, that less developed countries should be assisted in building capacity in order to reap the significant benefits from GM varieties. Other potential gains for developing countries include revenues generated from the use of their biodiversity by researchers from developed countries.

Social-development interests argue that there are problems with this global-welfare promise (Shiva, 1985; Rural Advancement Fund International, 1998; Kneen, 1999). First, the leadership of the private sector in agricultural biotechnology means that much of the knowledge is proprietary and will not be shared without compensation, a price that many producers in less developed countries cannot afford. As a result, the overwhelming majority of GM crops have been developed to meet the needs of developed-country producers – those who can afford to pay for the technology.

Secondly, most developing countries lack the current scientific capacity necessary to be involved in GM-crop development. In 1998, the World Bank reported that the gap between developed and developing countries in terms of knowledge capacity exceeded the income gap (World Bank, 1998). This is not just an issue of technology transfer, because these countries tend to lack the educational infrastructure to adapt scientifically, let alone the financial infrastructure to support competitive, private agricultural biotechnology ventures. The dynamic scale effects of increased potential for new GM varieties combined with decreased time for product development suggest that the capacity gap will only increase. Further, the capacity gap limits the possibility of developed countries relocating their research and development efforts to a developing country in order to share the knowledge. In fact, an alarming development from the perspective of global economic development is the potential import substitution of traditional exports from less developed countries. For instance, recent research efforts on canola (rape-seed) in both the USA and Canada have focused on modifying the oil content of the seed to express the characteristics of palm oil. However, several developing nations rely heavily upon palm-oil exports as a source of foreign currency.

Indeed, it appears that the potential economic-development problems with GM crops for less developed countries are institutional problems of access to and the distribution of technology. As agriculture increasingly becomes a high-technology industrial sector, driven by private firms, the competitive disadvantage of developing countries is likely to grow, with a dramatically decreasing likelihood for these countries to catch up. Since knowledge is the key, industrial imitation is not possible, as it was in other high-technology sectors of the past. In order to prevent such developments, institutional reform in both developed and developing countries, focused on capacity building, technology transfer and the targeting of innovations to the production problems in less developed countries must be undertaken (Rural Advancement Fund International, 1998; Falconi, 1999).

Accordingly, the currently commercialized production-trait GM crops contravene the belief system of social-development organizations, because they are being developed by multinational corporations driven by the profit motive with little regard for the socio-economic needs of developing-country farmers, who may be displaced by the technology. Hence, social-development organizations support regulatory development and integration policies ensuring technological precaution.

Notable social-development organizations critical of GM crops include: Council for Responsible Genetics, USA (http://www.gene-watch.org/); International Center for Trade and Sustainable Development, USA (http://www.ictsd.org); Foundation for Economic Trends, USA (http://www.biotechcentury.org/); Gen-Ethic Network (http://www.geneethics.org/); Christian Aid (http://www.christian-aid.org.uk/); World Development Movement (http://www.wdm.org.uk/); Rural Advancement Fund International (http://www.rafi.org/); Canadian Institute for Environmental Law and Policy (http://www.cielap.org/); Third World Network (http://www.twnside.org.sg/); Edmonds Institute, USA (http://www.edmonds-institute.org/); and the Latin American Consortium on Agroecology and Development (http://www.cnr.berkeley.edu/~agroeco3/clades).

4.2 Social Regulatory Integration: Case-study of the Cartagena Protocol on Biosafety

The Cartagena Protocol on Biosafety (BSP) represents an international multilateral environmental agreement on the transboundary movement of biotechnology in pursuit of deeper social integration.[2] It is important to assess for several reasons. First, the protocol is widely supported by social-interest organizations, because it is based on the social-rationality approach to regulatory development and integration. Secondly, the EU has stated that rules and guidelines established

[2]Secretariat of the Convention on Biological Diversity (http://www.biodiv.org/biosafety/).

under the protocol will be the foundation of its regulatory integration strategy dealing with biotechnology products (*AgraFood Biotech*, 1999a). Hence, it is a proxy for the EU's trade-policy position on GM crops. Thirdly, it effectively illustrates the conflict between trade-orientated economic integration and deeper social integration.

The BSP negotiations were an international effort, under the auspices of the United Nations Environment Programme (UNEP) 1992 CBD. UNEP was established in 1972 by the General Assembly as the UN's official environmental agency. Its mandate was sustainable development: to safeguard and enhance the environment for present and future generations. UNEP is involved in both technical scientific research on the environment and in reconciling the global objectives of environmental protection with other objectives, such as trade and economic development. For instance, UNEP is involved in negotiations to establish international environmental law through MEAs, such as the Montreal Protocol and the Basle Convention on the Control of Transboundary Movements of Hazardous Wastes and their Disposal.

Perhaps the most ambitious initiative of UNEP is the CBD and the subsequent negotiations to create a BSP. The CBD was the culmination of a decade-long effort, begun at the Third World Congress on National Parks and Protected Areas in Bali, Indonesia, in 1982 (Swanson, 1997). The objective of the CBD was to develop an international convention to commit the global community to conserving and protecting biodiversity. In June 1992, the CBD was included as Agenda 21 of the United Nations Conference on Environment and Development in Rio de Janeiro, the Earth Summit, and was signed by participating countries at the Conference. The BSP was proposed as an initiative to regulate the transboundary movement of living products of modern biotechnology in order to protect biodiversity. Negotiations were concluded in January 2000, after being suspended in February 1999 because of significant obstacles. Essentially, the obstacles to the BSP are associated with the conflict between social interests, supporting environmental objectives, and economic interests, supporting trade objectives; the BSP is driven from the environmental side and yet has significant potential trade implications.

In order to protect biodiversity, the initial scope of the BSP was to develop legally binding international rules governing the testing, importation and exportation (transboundary movement), deliberate release and commercial use of living modified organisms (LMOs). It included the advance informed agreement (AIA) principle, which meant that the Party of Import would be notified prior to a shipment of LMOs, so that it could perform a risk assessment in order to identify any potential risk(s) to regional biodiversity. The Party of Import, upon completion of the risk assessment, could allow or restrict the importation of the LMO because of identified risk(s) to biodiversity.

Although the BSP is not explicitly intended to be a trade agreement, the fact that its scope includes export and import activities makes it an implicit or *de facto* trade agreement associated with the international trade of GM products. A successful protocol has the potential to positively influence international trade in three significant ways: first, through increased trade transparency, according to the use of the AIA principle; secondly, through increased trade fairness, because the risk assessment procedures are intended to ensure that biodiversity risks from GM products, whether domestic or foreign, are assessed consistently, using credible procedures – that is, the use of science in performing risk assessments on products of biotechnology would be harmonized among all signatories; and thirdly, an international protocol could overcome the lack of domestic regulations in those countries with little or no experience with regulating GM products (Mulongoy, 1997). In this sense, the successful negotiation of the BSP can be interpreted as a potential win–win outcome. The global benefit, shared by all countries, is the overall conservation and protection of biodiversity. From an industry perspective, successful completion of the BSP has potential benefits for the further research, development, adoption and commercial use of GM products because it would potentially increase predictability and market-access opportunities.

In order to understand the current nature of the BSP, it is useful to understand the complicated negotiation procedure. As mentioned, the CBD Secretariat – located in Montreal, Canada – administers the BSP. Over 120 countries were involved in the seven BSP negotiations, held in: July 1996, Aarhus, Denmark; May 1997, Montreal, Canada; October 1997, Montreal, Canada; February 1998, Montreal, Canada; August 1998, Montreal, Canada; February 1999, Cartagena, Colombia (suspended talks); and January 2000, Montreal, Canada.

The negotiations began with the discussion of general issues, including who should be involved in the negotiating sessions. As an international governmental arrangement, the main actors would be national governments; however, the BSP negotiations also included a broad range of industry representatives and environmental non-governmental organizations (ENGOs). In this sense, the BSP employed a high level of openness from the beginning. A request for draft protocol submissions by October 1996 was made. Ethiopia submitted a draft protocol on behalf of the African delegation – written by the Third World Network (TWN) – in October 1996. This draft protocol, considered as representative of the views of many developing countries, used as a framework the Basle Convention on the Control of Transboundary Movements of Hazardous Wastes and their Disposal (McDonald, 1997). As a result, the draft protocol treated shipments of LMOs with the same degree of prescriptive regulation as shipments of toxic or nuclear waste. Further, this draft protocol placed enormous burdens upon the Party of Export and the exporter to ensure biosafety and to gain approval before any shipment of LMOs.

In response to the draft protocol submissions, the second negotiating session involved parties staking out their positions. The third session, in October 1997, was characterized by the emerging awareness of the agricultural-commodity trade issue and the potential impact of the protocol upon the international trade of products of modern biotechnology. Agricultural export countries reacted negatively to the Ethiopian draft protocol, highlighting the substantial differences of opinion between many developed and developing countries on what constitute LMOs and their risks. The fourth and fifth sessions primarily involved the edification of crucial definitions and issues, including the definition of LMOs, the roles of the Party of Export, the exporter, the importer and the Party of Import, the opportunity for exemptions and the scope of the AIA. Many of these issues remained unresolved.

The sixth negotiating session was to be followed by the Extraordinary Conference of the Parties to the Convention (ECOP), where the final draft BSP was to be presented for signing. However, on 24 February 1999, after it became clear that a final draft protocol was not going to be established, the decision was taken to push back the deadline for the final protocol for 18 months. The impasse emerged when the Miami group of countries (Canada, Australia, Argentina, Chile and Uruguay) rejected European efforts, supported by the other 140 negotiating countries, to extend the coverage of the Protocol in two directions: (i) to include risks to human health; and (ii) to include agri-food shipments intended for processing. The position of the Miami group was driven mostly by trade concerns. Specifically, the ambiguous interpretations of AIA and LMOs, along with the unclear provisions on labelling and liability laws, meant that there was significant uncertainty as to what trade impact the BSP might actually have on agri-food exports from countries where GM crops had been commercialized. For instance, a study of the potential trade impact – given the ambiguity – concluded that the impact on Canadian agricultural exports could range widely – from as low as C$6 million to as high as C$1200 million (Isaac and Phillips, 1999a). Until this severe uncertainty could be cleared up, major agricultural exporters simply could not support the protocol.

The seventh negotiating session, 24–28 January 2000 in Montreal, Canada resulted in the signing, by 140 countries, of the Cartagena Protocol on Biosafety. In its final form, the significant ambiguities concerning the Miami group were either dealt with or at least set aside for further consideration. For instance, the definition of LMO distinguishes between LMOs intended for environmental release and GM commodities not intended for environmental release. The former are subject to an AIA for the first-time shipment only. The protocol also specifies that there must be a clear scientific justification to ban either LMOs or AIA; however, it also allows a 'precautionary' ban, reflecting an environmental-type social interpretation of the precautionary principle, which can include socio-economic risks, such as impacts on local farmers (to be discussed in greater detail in Chapter 5). While GM commodities are not subject to an AIA, they do require mandatory labelling of GM material for the consumers' right to know. However, both the labelling rules and the liability rules are provisional, pending further negotiations.

In respect of the prospects for enhancing regulatory integration, the BSP is intended to create an international regulatory approach to the protection of biodiversity from LMOs, within the package of other international environmental protocols included in the CBD. In fact, attempts have been made to link the BSP to other UNEP international environmental regulations, including the International Technical Guidelines for Safety in Biotechnology, which are guidelines for the development of domestic regulations that deal with the safe handling and containment of GMOs within a country, and the United Nations (UN) Recommendations on the Transport of Dangerous Goods, which are recommendations for the development of domestic regulations pertaining to the transport of hazardous and toxic materials. Notable for its absence is any effort to link the potential BSP with international trade agreements. In this sense, the BSP is an environmental-protection agreement dominated by the social-rationality ethos, congruent with the broader socio-economic concerns that the WTO framework fails to address. Proposed regulations are process- or technology-based and the integration strategy is one of regulatory coordination.

Several crucial issues remain unresolved, which dampens enthusiasm for the BSP. As mentioned, and unlike the IPPC, there is no link to any international trade agreements, even though it is a *de facto* trade agreement, and there is no institutional mechanism for dispute settlement. Secondly, the USA is not a signatory. Although not an official negotiating party (the US Congress has not ratified the 1992 CBD), the USA remains the world leader in biotechnology research and plays an influential role in the negotiations. Whether the US position assumes a cautionary acceptance or an outright opposition to the BSP can have a vital influence on those signatories to the protocol which rely heavily upon market access to the USA, such as Canada. Thirdly, it appears that it will be at least 2008 before the BSP is put into effect. Negotiations on the provisional labelling rules continued until 2002, while those for the provisional liability laws continue until 2004. Once settled, the BSP must be ratified by at least 50 signatories and then transposed into national laws. Of course, during this time, technological innovation in agricultural biotechnology will continue, perhaps making the BSP obsolete as an international regulatory development and integration agreement, even before it becomes ratified as an international treaty.

In fact, in many respects, it appears that the BSP came about more as a political compromise than as an actual attempt to establish an institutional structure for GM-crop regulations. The most contentious issues remain 'provisional' (unresolved) and the time-consuming ratification and implementation process could ensure its obsolescence. Its major contribution was the illusion that an international agreement blending the scientific- and social-rationality approaches and the economic and social perspectives had been achieved.

4.3 Conclusions

From a regulatory development and integration perspective, social-interest organizations play a vital role in finding the regulatory middle, as their extreme concerns counteract the extreme global-welfare promises of the life-sciences companies. In general, most of the social-interest organizations are dissatisfied with GM crops, because they contravene a particular belief system. What is clear from the above assessment is that most of the opposition to GM crops is the result of how the technology has been applied, by whom and for whom, providing a foothold for a vast coalition of concerned social-interest organizations. Indeed, an interesting aspect of the European anti-GM campaign is that, while GM crops are not commercially grown in Europe, environmentally unsustainable agricultural practices supported by the Common Agricultural Policy (CAP), which account for over 95% of EU production, continue without the concerted negative campaign targeted at GM crops. Perhaps this is the case because GM crops provide a convergence of fear of the unknown, associated with their credence factors and a perception of an invasion of foreign technology. Environmental organizations, such as Greenpeace and Friends of the Earth, would be unable to generate as much public exposure and support if they were taking direct action against the commercial crops of European farmers, instead of the field trials of multinational life-sciences firms.

In some cases, the normative beliefs are so strong that the organizations simply will not compromise and may be willing to take illegal action. Consider, for instance, environmental organizations. There has always been a minority element of eco-warriors, whose extremism was not supported by all and could often be linked to a more holistic rejection of capitalism, technology and modern society. However, in the case of GM crops, illegal action has shifted to more mainstream organizations. For instance, in the UK in 1998, field trials of GM crops were destroyed by the more extreme group GenetiX Snowball. By 1999,

Friends of the Earth and Greenpeace were involved. The Director of Greenpeace, UK, Lord Melchett, justified the illegal action by essentially claiming that 10 years of legal protests did not result in regulatory policies adopting Greenpeace's belief system. This is a dangerous challenge to pluralism, which necessarily demands compromise and concession. It sets a dangerous precedent to justify illegal action because of the dissatisfaction that normative beliefs are not met in their entirety in the regulatory system. An example of the unwillingness or inability of Greenpeace to compromise was revealed in late 1999 at the Greenpeace Business Conference in London. Lord Melchett issued a challenge to the then Chairman and CEO of Monsanto, Bob Shapiro:

> If Monsanto will stop developing GM crops, get out of producing pesticides and reject the idea of patenting life forms, Greenpeace will work enthusiastically with you . . . to produce a new Monsanto . . . We could create a genuine life sciences company based on ecological, organic, holistic principles.
>
> (Greenpeace, 1999)

In essence, this challenge is akin to asking Greenpeace to stop environmental advocacy.

Moreover, it is also reasonable to suggest that social-interest organizations pursue their own self-interest in a manner that cannot be disentangled from the self-interest pursued by GM-crop developers, implying that caution must be exercised in assuming the altruism of such organizations. These organizations require membership to survive.[3] In this sense, many of them are in competition with one another for members, and championing emotive issues through dramatic campaigns, or 'picking good fights', is an important method for sustaining public awareness of the organizations (Bode, 1998). Given the range of consumer concerns and the large and growing information gap between producers and consumers, there can be no doubt that GM crops is a good fight to pick, because particularly emotive arguments can be easily developed.

Finally, the rise of civil-society advocacy at the international level against the globalization of the international trading system has, in fact, created its own form of NGO globalization. In this case, an undemocratic and unaccountable foreign influence reaches deep into national policy competence and constrains the regulatory development and integration strategies that a nation can choose under the threat of mass protest. Indeed, this NGO globalization creates a domestic democratic deficit not unlike the democratic deficit that the

[3]It has been noted that in 1991 membership in Greenpeace was 4.8 million and in 1999 it had fallen to 2.5 million (*Economist*, 1998). The same article also notes that Greenpeace has considered the development of a Greenpeace eco-label.

organizations claim exists in the international trading regime. Given that most of these social-interest organizations would like to see the establishment of a strong international institution with as much power as, if not more power than, the WTO to enforce the social-rationality approach to environmental protection, it can hardly be said that they are anti-globalization.

This raises three important questions. First, despite the very public opposition to GM crops by many social organizations, do they really have significant influence in the regulatory development and integration strategies chosen by a jurisdiction? The answer is yes, and policy-makers ignore them at their peril. In the UK, Monsanto has admitted that the anti-GM campaign, which has been mostly portrayed as an anti-Monsanto campaign, has been damaging to its business reputation and its share price (*Independent*, 1999q). As a result, the Monsanto President and Chairman made efforts to establish a dialogue with the anti-GM campaigners. The interesting problem will be the willingness and ability of the social-interest organizations to compromise their belief systems. Have they created such a public resistance that, when they realize there are benefits from GM crops, they will be unable to compromise? For instance, after a sustained campaign against the 'technology' of GM crops, Greenpeace had to soften its position in respect of the development of the GM vitamin A rice and to admit that certain applications were essentially positive (*Telegraph*, 10 February 2001). This was the first instance in which Greenpeace publicly distinguished between the technology *per se* and the application, management and distribution of the technology. Moreover, this softening of the Greenpeace position supports the argument for why regulatory development should be science-based and rules-based according to actual risk, rather than based on a reactive approach to assuaging public perceptions, unsubstantiated concerns and unsubstantiated risks.

Secondly, given the fact that in most cases it is not the technology but the application, management and distribution of the technology that are of concern, what would have happened if the first GM crops were developed by supporters of organic farming in a public research institute designed to employ the technology to assist the sustainable development needs of less developed countries, rather than by multinational corporations targeting intensive agricultural farmers in developed countries? It is a question worth considering. Given the application focus of these interests, it is reasonable to argue that, if the initial commercializations were different, so too would be the reaction of many of the current opponents. As mentioned previously, a prototype credit card made by Monsanto from plant fibres was initially regarded as an environmentally sustainable solution to the problem of petroleum-based plastics.

Thirdly, although there has been a considerable range of organizations united in opposition to GM crops, is this unity of opposition sustainable in the medium to long term? For instance, consumer organizations are in support of higher domestic food-safety and environmental-protection standards. Social-development organizations want better access to developed markets for the exports from less developed countries. Yet setting income-elastic food-safety and environmental protection standards in developed markets could prohibitively bar the exports of these countries. The incompatibility of all of these organizations lends credence to the argument that they have chosen GM crops because of its easily exploitable emotive characteristics.

In summary, the examination of the opposition to GM crops from consumer, environmental and social-development organizations reveals several interesting results. First, it is not the GM crops *per se* that are the source of dissatisfaction. Rather, they are rejected because the initially commercialized applications are produced for an agronomic system and by multinational corporations who play by the WTO's trade rules, which contravene the belief systems of the social-interest organizations. Secondly, in the short to medium term, the opposition to GM crops is formidable and neither government regulators nor industry can ignore it. Thirdly, the current unity of opposition to GM crops among various organizations with incompatible objectives does not appear sustainable in the long term as the GM-crop applications move from a homogeneous focus on input traits to a focus on output traits and novel products with a different balance of benefits and burdens for the consumer. Fourthly, it appears that the opposition to GM crops, resulting in calls for greater regulatory oversight and weaker intellectual-property protection is, in fact, having the effect of increasing the concentration of agricultural life-science firms – that is, multinational firms are forced into greater concentration in order to deal with the effects of stringent regulatory oversight and unclear intellectual-property protection rules.

Therefore, the development and integration strategies for GM regulations must focus to some degree on the application, management and distribution concerns advocated by the social-interest organizations regardless of whether or not there is a scientific justification for the concerns. The challenge, of course, is identifying the appropriate degree of focus.

5 Regulatory Development and Integration

Given that economic and social interests support very divergent approaches to regulatory development and regulatory integration, establishing a regulatory development and integration approach must be understood as a political exercise of balancing technological progress with technological precaution. The objective of this chapter is to analyse the development of genetically modified (GM)-crop regulations in order to identify the sources of regulatory instability and the prospects and limits of regulatory integration. Essentially, in this chapter, a conceptual model for determining a jurisdiction's regulatory framework, specific GM-crop regulations and, consequently, its regulatory integration strategy is developed.

5.1 Regulating Technology: the Risk-analysis Framework

At the core of the debates on regulating GM crops is the issue of regulating technology – a challenging public-policy problem because technological innovation, while promising opportunity, has always encountered public anxiety about its risks. Such anxiety has always been associated with agricultural innovations:

> We have already advanced our knowledge of genetics to the point where we can manipulate life in a way never intended by nature. We must proceed with utmost caution in the application of this new found knowledge.
>
> (Burbank, 1906, in Bruhn, 1993)

This quote captures many of the current concerns about GM crops. It was written, however, in 1906 and it refers to the techniques of traditional plant breeding. Interestingly, the techniques that many now consider as the 'safe old methods' were met with anxiety and hostility similar to those currently facing GM crops.

Technological innovation is a dynamic process, where research and development build both capacity and the knowledge base, which in turn increase the potential for further innovations and for specific commercial applications (Phillips and Khachatourians, 2001). Yet this simultaneously increases the information gap between producers and consumers and exacerbates the credence nature of advanced-technology products such as GM crops. The increasing sophistication and credence nature of the technology decrease the ability of consumers to understand the actual risks and to act rationally on their own. Increasingly, consumers must rely upon others, such as industry, government regulators or a third party, to address the credence concerns on their behalf.

In this sense, the development of regulations to deal with GM-crop technologies must focus on two broad objectives. First, regulation must strike a political balance between technological progress and technological precaution, which is essentially a balance between the benefits and the risks of new technology. As there is no such thing as certainty in assessing the benefits or the risks of new technologies, it is the acceptability of the relative benefit/risk trade-off that matters. In achieving this, regulations cannot be expected to build consensus among normative and perhaps quite divergent views on technology (van den Daele *et al.*, 1997). The best that regulatory approaches can hope to achieve is an economically and socially acceptable balance, not a risk-free balance.

Secondly, regulations must be dynamic and capable of keeping pace with the commercial applications of new technologies. This means that, as GM-crop applications deepen and widen, regulatory foresight is required in order to anticipate issues arising from new production and output of improved GM crops and from novel bioengineered products, so that adequate risk-assessment capacity is in place when the time comes. Of course, this puts pressure on the limited resources of government regulators, implying that regulatory resources must be strategically managed. It has been argued that the regulatory goal should not be to develop a comprehensive regulatory approach of complete command and control, because it will quickly become obsolete. Instead, the goal should be to develop a dynamic management approach capable of steering 'the matured technology in beneficial directions' (Fincham and Ravetz, 1991).

The risk-analysis framework (RAF) is widely considered within both the economic and the social perspectives to be an effective framework for simultaneously addressing the complexities and the limitations facing regulatory development as a result of sophisticated technologies such as GM crops. As previously discussed, the RAF was first outlined in 1983 as being comprised of three parts: risk assessment; risk management; and risk communication (National Academy of Sciences, 1983).

Risk assessment has the goal of developing, to the extent possible, objective, neutral, transparent and consensual information about the risks without normative influence. Risk assessment involves hazard identification, exposure assessment and risk characterization according to a sound scientific basis, using accepted analytical methods and statistical-inference techniques. The information gathered in the risk-assessment stage is then used to inform the risk-management process.

The goal of risk management is risk prevention and risk reduction. This involves incorporating the objective risk information into a regulatory response. As risk management is where the regulatory position is adopted, this stage must also address the broader economic, political, social, moral and ethical factors beyond a scientific basis. In this sense, risk management explicitly requires compromise and concession, as it balances the rights and interests of all stakeholders in order to establish an acceptable regulatory framework.

The Redbook specified that risk-assessment techniques should be institutionally separated from risk-management procedures. Essentially, it is vital to disentangle the objective, scientific risk assessment from the normative public response to risk.

Risk communication is a two-way flow of information between the risk-assessment and risk-management stages, as well as between regulators, industry and the public. While the scientific risk-assessment information must flow to the regulatory development process in risk management, it is also necessary for the social parameters of risk management to flow back into the risk-assessment techniques (MacKenzie, 1993). As was shown in Chapter 2, there has been a failure of risk communication in respect of GM crops. Much of the information provided for consumers has been either the industry's global-welfare promise or from the highly critical environmental lobby. Such a polarity of information has left many consumers confused and, given their propensity to trust third-party non-governmental organizations, many consumers have an unsubstantiated negative perception of GM crops. In this sense, the goals of risk communication would be to reduce the credence factors and the information gap and to increase confidence in the regulatory framework through greater transparency.

A vital contribution of the Redbook's RAF was the integration of scientific analytical procedures into the regulatory development process. There are two important reasons why science matters. First, scientific analysis has long striven to disentangle normative values from scientific discovery. Although a complete separation has not occurred – and will never occur – there is a level of universality in 'science' that is not present in social beliefs, morals and ethics. As regulations apply across a wide range of social beliefs, morals and ethics, they should endeavour to be as normative-free as possible. For instance, food-safety regulations should be focused on human safety, equally applicable to all citizens, and not the social values of some, while contravening the values of others.

Secondly, while scientific disagreement can and frequently does occur, the analytical methodology of science permits debate on disagreement subject to relatively accepted rules of evidence,[1] unlike normative disagreements which, based on beliefs, often have no apparent methodology or process for resolution. The RAF seeks to establish a sound scientific basis for risk assessment in order to develop, as far as possible, neutral and objective information to advise the risk-management process, while risk communication is intended to provide transparency for the entire regulatory development process.

Despite the widespread support for the RAF, the examination of economic and social interests in the previous chapters revealed that significant differences exist on how to develop regulations appropriately. In fact, there are seven framework debates associated with regulating technology that represent a conflict between the RAF supported by economic interests and that supported by social interests. Within the context of these framework debates are five specific debates on GM-crop regulatory principles, which are examined after the general debates. It will be shown that GM-crop regulations are unstable, as they shift within the risk-analysis parameters according to the competing influence of economic and social interests. As these debates fundamentally influence the path of regulatory development within a jurisdiction and the prospects for regulatory integration between jurisdictions, it is crucial to examine each one.

5.1.1 Scientific rationality vs. social rationality in risk assessment

In a general sense, there are two dominant approaches to the RAF consistent with the two perspectives: the scientific-rationality paradigm and the social-rationality paradigm. The type of rationality adopted essentially establishes the trajectory for the subsequent regulatory development. Indeed, this dichotomy is at the heart of the conflict between the economic and the social perspectives. It is now possible to define this dichotomy in a regulatory context.

According to the scientific-rationality or traditional rational-thought perspective, it is possible to derive scientific facts that are

[1]See Chapter 7 for an illustrative example of how scientific rationality dealt with disagreement in the case of Dr Pustzai.

universal and that lead to certainty. In one approach, it has been argued that researchers can ask either scientific empirical questions or social normative questions (van den Daele *et al.*, 1997). Empirical questions can be assessed according to accepted analytical methods, with specified causal–consequence mechanisms to show risk. Such questions for GM crops would include whether a risk exists and how likely it is to happen. Empirical questions produce 'matters of knowledge', which can be based on consensus, while normative questions involve consideration of different systems of belief, not based on any analytical method. Social normative questions for GM crops would include whether the risk is acceptable. These questions are not prone to consensus and there is no obvious method for resolving conflicts among belief systems. In comparing the two, scientific rationality posits that empirical questions are logically prior to normative questions; hence, empirical questions can produce neutral, objective and universal facts, which should be the basis of risk-assessment procedures (Table 5.1).

On the other hand, the social-rationality paradigm rejects the idea of universal facts. Instead, empirical questions and answers are considered to be scientific myths, because facts cannot exist without value and social dimensions cannot be separated from science. The social-rationality paradigm suggests that neutral and objective risk assessment is a flawed concept. It challenges the scientific RAF on the basis that, in distilling risk down to a mechanistic causal–consequence model, the scientific RAF fails to address the complex interactions between technology and society (Giddens, 1994; Grove-White, 1997; Grove-White *et al.*, 1997; Sheppard, 1997).

The scientific- and social-rationality foundations for regulating technology can produce a fundamental conflict in regulatory development and integration – a conflict that is the basis for the remaining framework debates on the role of science in GM-crop regulations. On the one hand, economic interests view scientific risk assessment as a means of disentangling actual risk from perceived risk and from normative social preferences distorting the first-best regulatory policies. It is held that empirical analysis according to accepted methods and causal–consequence models provides a degree of universality for the

Table 5.1. Rationality of the risk-analysis framework.

Scientific rationality	Social rationality
Empirical questions	Normative questions
Facts or matters of knowledge	Values or social parameters
Technological progress	Technological precaution
Food-safety basis	Environmental-protection basis

risk assessments of GM crops, providing a certain, rules-based foundation for regulations and permitting technological progress. On the other hand, social interests view social risk assessment as a necessary requirement for establishing socially responsive regulations, permitting technological precaution. In this sense, they do not distinguish between actual and perceived risk, because perceived risk is held as an equally legitimate trigger for regulatory change and development. Given this difference, it is vital to recognize the rationality underlying the regulatory development process within a jurisdiction, as will be done in Chapters 6 and 7 in Part III.

5.1.2 Type of 'risk' targeted in risk assessment

Another important debate associated with the RAF involves the type of risk targeted, where three types may be identified: recognizable risks, hypothetical risks and speculative risks (van den Daele *et al.*, 1997). Recognizable risks can be identified through experience (data) and the application of accepted analytical methods, such as statistical inference and probability theory, and they include a clear causal–consequence mechanism. Hypothetical risks lack experience or data, but, with the help of assumptions and/or likelihood functions, they can be assessed within an accepted analytical method. Speculative risks lack experience, data, a causal–consequence mechanism and an accepted analytical method for assessment. They are logical possibilities – irrefutable, but also untestable.[2]

An added dimension to the type of risk is the risk consequence, where there are two types of consequences: dread and diffuse. Dread consequences are sudden and catastrophic, while diffuse consequences are long-run and cumulative but not catastrophic. It has been argued that dread consequences have important implications for risk-related behaviour. They can upwardly bias the perception of both consumers and regulators about possible recurrence (Camerer and Kunreuther, 1989) and they can create ambiguity-averse behaviour, where consumers and regulators overestimate the risk of products because of the credence factors (MacLaren, 1997). The most severe event in terms of creating ambiguity-averse behaviour is when speculative risks are combined with dread consequences.

New technology always faces either hypothetical or speculative risks, because by definition there is little experience with or few data

[2]For examples of statements of speculative risks of GM crops see the Consumers' Association (1997), Sheppard (1997) and the *Ecologist* (1998). Although only a brief list, these references clearly indicate the widespread use of speculative risks in respect of GM crops. In each, no recognizable or hypothetical risks are proposed; instead, they all refer to dread consequences of GM crops based on speculative risks.

available on the risks. The controversy arises over the extent to which GM crops should be assessed according to hypothetical or speculative risks (Table 5.2). Many supporters of a social-rationality paradigm have made dire predictions of 'genetic pollution' and ecological disaster because of the use of GM crops (Ho, 1998; Rifkin, 1998a,b). They appear to focus on speculative risks and argue on the basis of logical possibilities targeted to public concerns about the credence factors of GM crops and alluding to dread consequences. Due to these speculative risks, which cannot be refuted, they claim that a scientific RAF is insufficient in properly assessing the risks of GM crops. Without a sufficient RAF, GM crops must be banned.[3]

In contrast, supporters of scientific rationality may argue that the problem is not with the scientific RAF, but with the deliberately vague nature of speculative risks (Fincham and Ravetz, 1991; van den Daele *et al.*, 1997). They agree that speculative risks cannot be proved wrong, as 'it is scientifically impossible to prove the impossibility of an unwanted event; and this may seem to be what is demanded by protesters' (Fincham and Ravetz, 1991). Yet they counter that speculative risks cannot be proved right either. From an operational point of view, a regulatory risk assessment can only address a finite set of risks, despite the fact that there exist infinite logical possibilities. Moreover, there is a precautionary default built into a scientific risk assessment of hypothetical risks: if a risk is so poorly understood that no accepted analytical model can be developed, then no risk assessment can be done and the regulatory decision is made that there is too much uncertainty to proceed.

A social-rationality regulatory approach targeting speculative risks would result in no new technology ever being approved, because speculative risks can always be found. But then a problem emerges, because an infinite number of speculative risks can also be identified for previously approved old technologies, such as traditional plant-breeding techniques, so must they too be rejected? Of course, they would not be because, due to experience and data, they are now associated with recognizable risks. Banning a technology because of speculative risks prohibits the shift from hypothetical to recognizable risks. Scientific rationality posits that speculative risk assessments are not an acceptable basis for regulatory development, and instead hypothetical risk assessments must be the basis for managing the application and distribution of new technologies. Accordingly, speculative risks are viewed simply as the safe harbour for those uninterested in a genuine dialogue about new technology and are deemed inappropriate, because regulations are about balancing the rights and interests of supporters and critics, requiring dialogue, debate and compromise.

[3]Also the National Consumers Council (1998) defines hypothetical risk as 'risk' and speculative risk as 'true uncertainty' and then argues that risk assessment is problematic because it does not take account of 'true uncertainty'.

Table 5.2. Type of risk in the risk-analysis framework.

Scientific rationality	Social rationality
Recognizable	Recognizable
Data, accepted analytical models	Data, accepted analytical models
Hypothetical	Hypothetical
No data, but accepted analytical models	No data, but accepted analytical models
	Speculative
	No data, no accepted analytical models
	Logical possibilities

5.1.3 Substantial equivalence in risk assessment

The equivalence of GM and non-GM crops is a very important debate associated with GM-crop regulations based on the RAF. Substantial equivalence is a 'gateway' regulatory principle used to determine the novelty, which is the key aspect in determining the regulatory path that a product must follow. The basis of this principle is the notion that it is products or the application of technology that should be the regulatory focus, not the use of the technology or the process *per se*. Operationally, this implies that, regardless of the differences in the processing and production methods (PPMs), the risks from a product are substantially equivalent to the risks from another 'like' product. It is important to note that, as a gateway regulatory principle, substantial equivalence cannot be used to assess risk; it is simply used to assess whether the risk of a new product is equivalent to the risk of a conventional product or not. If substantial equivalence has been found between, say, GM crops and non-GM crops, then the only interpretation to be made is that the risks from the GM crops are not greater than the risks from the non-GM crops. Consequently, the GM crops would be regulated according to the same standards as the non-GM crops.

The substantial-equivalence principle, in respect of GM crops, is applied in the following manner.[4] The first step is to establish an equivalence baseline from a range of conventional varieties of the same crop. Conventional varieties are examined across four broad categories, each with its own subcategories: (i) nutritional and natural composition; (ii) the level of macronutrients (e.g. protein, fat, carbohydrates, dietary fibre); (iii) level of vitamins (e.g. C, B_6, thiamine, niacin, folic acid, riboflavin); and (iv) the level of minerals (e.g. calcium, copper, iron, iodine, magnesium, phosphorus, potassium, sodium, zinc). The result is a baseline set of ranges for all the categories into which conventional crops fall. Then the GM variety is

[4]Summarized from: Gasson (1999), Kearns and Mayers (1999), Millstone *et al.* (1999) and McHughen (2000).

assessed across the same categories and subcategories and a profile is assembled. This profile is then compared with the baseline ranges, and values outside the conventional range are sought. If no values are outside the conventional range, then the GM crop is considered substantially equivalent and would be subject to the same regulations as the conventional crop. At the other extreme, if most values are outside the conventional range, then the GM crop is considered to be novel – that is, it would have values never before characterized or approved for the market. Consequently, the GM crop would face novelty regulations more stringent and precautious than conventional regulations.

What if only some of the values were outside the conventional range, while the remaining values were well inside? In this case, each of the values outside the normal range would be individually assessed and for each it would be determined why is it outside the range and whether this creates a safety concern. It may be outside, say, because of a higher level of insect resistance, but, if this is the value that was actually the reason for the genetic modification in the first instance, then the only test that would be required is for the safety of this outlying value. Of course, if a value were outside the conventional range and there were no explanation why, then the default would be to a precautionary approach, probably resulting in a determination of novelty, not substantial equivalence.

Economic interests, supporting scientific rationality, are in favour of the substantial-equivalence principle, because they claim that the key issue is not the application of modern biotechnology *per se*, but the novelty of the resultant crop. In support of this view, the United Nation's World Health Organization (WHO) published a report on the principle of substantial equivalence in safety evaluations of food products derived from GM crops (WHO/FAO, 1991). This report concluded that GM crops may be substantially equivalent to non-GM crops for risk-assessment purposes and that novelty was the justification for new regulatory oversight, not simply the use of genetic modification. According to this perspective, most production-trait GM crops and some output-trait GM crops are substantially equivalent, but bioengineered products will not be substantially equivalent, because they are, by definition, all novel. Substantial equivalence is a well-established principle in international regulations, supported by institutions such as the WHO, the Food and Agricultural Organization, the Codex and the International Plant Protection Convention (IPPC), as well as the Organization for Economic Cooperation and Development.

Social-rationality advocates do not accept substantial equivalence as a gateway regulatory principle, because they claim that it is the use of modern biotechnology that changes the fundamental nature of GM crops and incurs special risks not evidenced in conventional plant breeding. This is a process or technology focus, where no GM crops would be considered substantially equivalent even if all of the values fell within the conventional range (Fig. 5.1).

Substantial equivalence may be assessed over both scientific factors (van den Daele *et al.*, 1997) and normative factors (Nuffield Institute, 1999). Two popular arguments in support of the social-rationality assertion that GM crops are not substantially equivalent are that it is wrong to engineer living things and that it is wrong to transfer genes between species. Both of these arguments are biocentric ethical or moral arguments based on the idea that nature has boundaries and it is not right for humans to exceed those boundaries, regardless of the technological capabilities. With respect to the first argument, the notion of engineering plants cannot be limited to GM crops only. To reject GM crops on the basis that they are engineered requires a rejection of all agricultural crops, including organic varieties, because, it is argued, it is impossible to disentangle the engineering of GM crops from the engineering of non-GM agricultural crops (Nuffield Institute, 1999). Consider that GM crops are modified varieties of conventional crops already engineered and made unnatural by generations of traditional plant breeding. Similarly to all conventional crops, GM crops have been developed to fit within an 'unnatural' agricultural production system, not to grow in the wild, and to express desirable traits for human use, where these traits are not designed to confer any natural advantage on these crops relative to wild plants. It has been noted that 'by itself genetic modification does not normally confer qualities that will make an organism more harmful to mankind or the environment' (Fincham and Ravetz, 1991). In other words, to argue that it is wrong to engineer living things and reject GM crops on this basis requires a similar rejection of all agricultural crops, which have also been engineered to grow in an unnatural, human-made environment in order to meet human needs.

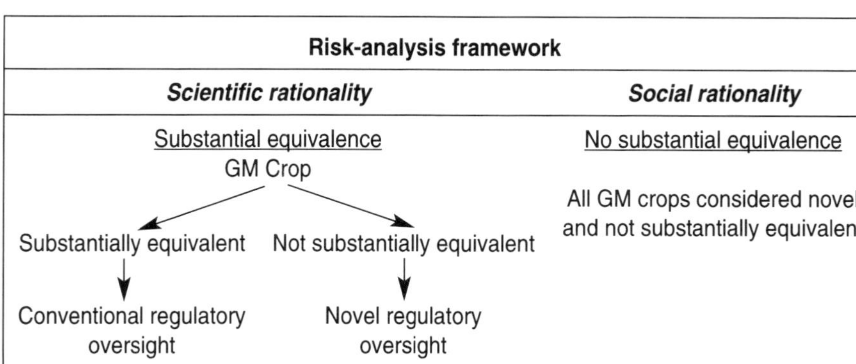

Fig. 5.1. Substantial equivalence in the risk-analysis framework.

Similarly, it is not a legitimate argument to reject GM crops because they are harmful to biodiversity (van den Daele *et al.*, 1997). By definition, all agricultural production is harmful to biodiversity. It must be accepted that production takes priority over biodiversity on agricultural land. The objective, given the priority of production, must be to minimize the impact of agricultural production upon biodiversity and there is no reason why GM crops cannot assist in minimizing the impact. For instance, an important example of the differing scientific- and social-rationality views on the substantial-equivalence principle concerns the relationship between GM crops and organic crop varieties. A common position is that organic produce provides consumers with a GM-free alternative. However, the debate arises because many seed developers reject the notion that GM seeds cannot be used in an organic production system. They argue that organic refers to a more labour-intensive production process, eliminating the use of artificial fertilizers, chemicals and pesticide residues and replacing them with a more integrated crop-rotation and pest- and weed-management system. In this sense, organic is about the production method, and GM crops developed to be congruent with such a production method are substantially equivalent to conventional organic varieties. Critics of the substantial-equivalence principle argue that it is the use of GM techniques that fundamentally changes the nature of GM crops and makes them incompatible with organic systems.

With respect to the second argument, from a scientific perspective, gene transfer and modification even across the species barrier is a naturally occurring phenomenon, not specific to genetic-modification techniques.[5] Indeed, genetic drift has always occurred between non-GM crops, wild relatives, other vegetation and soil microorganisms. In fact, it may be argued that GM crops involve more precise and specific modifications, which may be more carefully monitored, rather than the unmonitored random type that occur naturally, giving the plant breeder greater control than previously possible.

It would seem, given the above arguments, that from a regulatory perspective the key issue is not the genetic modification *per se*, but the 'artificialness' of the modification, such as the transfer of the antifreeze gene of arctic fish into crops to protect them from freezing. This is an application-specific concern focused on novelty, not a process-based concern associated with the use of the technology. This implies that, for risk-assessment purposes, some GM crops may be considered as substantially equivalent and regulated accordingly, provided they are not novel.

[5]Ho (1998) notes that viruses have the natural ability to transfer genes between unrelated/sexually incompatible species; therefore, transgenic modification cannot be seen as unnatural.

5.1.4 Science vs. other legitimate factors in risk assessment

There is debate over whether risk assessment should only consider science-based, empirical definitions of safety or should also involve broader socio-economic factors. From a food-safety perspective, there are four different regulatory targets or hurdles that may be employed in assessing the 'risk' or safety of foods derived from GM crops in order to determine market approval. The first target is food safety defined as the absence of short-term illness immediately after consumption – the classic food-poisoning case. The second target is long-term human health, such as the cumulative impacts of consumption over time. The third target involves assessing the quality or efficacy of the food product relative to other products in the same-use category. The fourth target includes assessments of the broader socio-economic impacts of the use of GM foods in the food supply. This includes, for instance, assessing the 'risks' of GM-crop production for the welfare of farmers who produce them and for those who produce complements or alternatives (Fig. 5.2). For example, in 1999, Thailand imposed a provisional import ban on GM material, pending proof of its social benefit, not just its safety. It was argued by Thailand's International Economic Relations Policy Committee that imports should not resume 'until the public accepts that GM food provides more benefits than drawbacks' (*World Food Law Monthly*, 1999a).

Most foods are analysed according to the first target – a scientific assessment of their potential to cause short-term illness only. Fast foods and junk foods with recognizable risks of long-term health implications are approved as safe and are widely available in the marketplace. In approving such foods, little consideration is given to the three other regulatory hurdles of long-term health consequences, food quality and the socio-economic impacts upon, for example, the health-care system or worker absenteeism because of chronically unfit consumers.

Social-rationality supporters generally argue that GM crops should

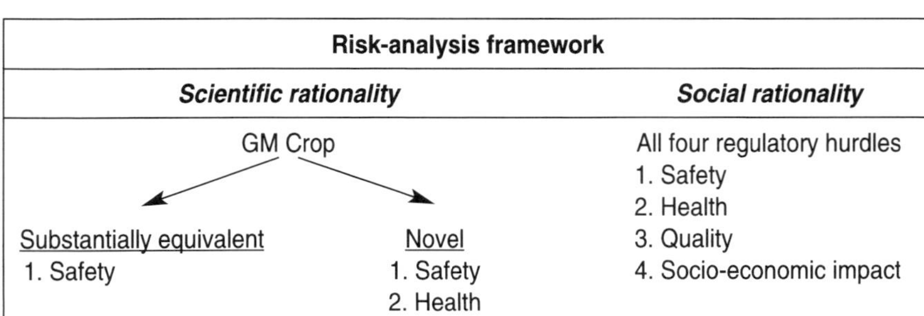

Fig. 5.2. Regulatory hurdles in the risk-analysis framework.

be assessed according to all four targets or regulatory hurdles. This position follows from both the focus on speculative risks and the rejection of the substantial-equivalence principle. The focus on speculative risks allows supporters of social rationality to consider risks that may not be verifiable according to a hypothetical risk assessment. Similarly, the rejection of the substantial-equivalence principle means that, according to the social-rationality perspective, all GM products are novel because of the technology used and, hence, subject to regulatory oversight across all four regulatory hurdles.

Scientific-rationality supporters reject this argument for three reasons. First, it is inconsistent with the traditional science-based approach to food-safety assessments. For instance, the 1995 amendments to the Codex procedural manual supported only the first regulatory hurdle (CAC, 1995). Secondly, there is no justification for assessing GM crops any differently from conventional crops in the food supply if they are considered to be substantially equivalent. Thirdly, even hypothetical risks associated with the safety of foods from GM crops have not been shown. In fact, a paradoxical development in the UK is the removal of GM ingredients from the products of several fast-food restaurant chains. No hypothetical risks to human safety and health have been shown from GM crops and yet there is considerable information on the recognizable risks from fast food. It appears that this development is simply not about food safety.

Similarly to the discussion on food safety, hypothetical risks to environmental biodiversity from GM crops may be analysed according to an empirical scientific assessment or according to a broader socio-economic assessment of other legitimate factors. For instance, the UK Economic and Social Research Council argued:

> [c]urrent approaches to environmental risk assessment exclude wider dimensions of risk, such as justification, need, benefit, the context of use, and public confidence in the trust worthiness of regulatory provisions and institutions themselves . . . the mechanisms in question are not appropriate for addressing such issues adequately and there are no other fora for addressing these dimensions.
>
> (Mayer *et al.*, 1996)[6]

Again, this would include, for instance, an assessment of the impact of GM-crop production upon the social welfare of other economic sectors both complementary to and competing with GM crops. A serious problem arises when a critic's dissatisfaction with 'need' and 'justification' is manifested through speculative, unsubstantiated and spurious warnings or dire predictions on the safety of GM crops.

[6]Also see Sheppard (1997), who argues that 'a re-evaluation of the risks, need and justification for GM foods, one that extends beyond what is normally thought of as "risk assessment" is urgently required'.

In short, scientific rationality supports the first hurdle for regulatory oversight when GM crops are deemed substantially equivalent and the first and second hurdles for novel GM crops, while social rationality supports all four regulatory hurdles for all GM crops.

5.1.5 Burden of proof in risk assessment

The traditional regulatory approach to new technology, supported by the economic perspective, generally reflects a scientific-rationality approach, where new technological innovations are innocent until proved guilty – that is, unless an innovation is proved to be harmful, there is no justification for a restriction or a ban on the technology.

To illustrate the issue of burden of proof in risk assessment consider statistical-probability theory, where there are two types of risk or error to consider. With respect to GM crops, a Type I risk would be the risk that the crops are rejected as unsafe when in fact they are safe, while a Type II risk would be the risk that the crops are accepted as safe when in fact they are unsafe. The controversy arises because Type I and II risks cannot be simultaneously minimized and, instead, if one type of risk or error is decreased, the other is increased (Greene, 1997). The debate is how to decide between the two.

The traditional burden of proof minimizes the Type I risk (rejected when safe) and supporters argue that this paradigm has been proved effective at ensuring an acceptable level of safety of innovative products without discouraging further innovation. Generally, this requires the assumption that firms actively minimize safety risks while pursuing innovation. It has been argued that firms are both able and willing to ensure both food and environmental safety, because a crisis is a commercial disaster (Spriggs and Isaac, 2001). According to this view, despite the claims of 'reductionist' science or captured scientists, the market is working in internalizing the costs of risk, because producers who minimize commercial risk are simultaneously minimizing food- and environmental-safety risks. Indeed, an important aspect of ensuring that firms

Table 5.3. Burden of proof in the risk-analysis framework.

	Safe	Unsafe
Accepted	Proper outcome	Type II risk Social rationality minimizes – guilty until proved innocent
Rejected	Type I risk Scientific rationality minimizes – innocent until proved guilty	Proper outcome

do internalize food- and environmental-safety risks is to have stringent product-liability laws to compel firms to ensure safety (Table 5.3).

The traditional burden of proof has important commercial advantages. It enhances regulatory stability and certainty for firms, encouraging the adoption of new technologies. This allows for hypothetical risks to begin to move towards recognizable risks, increasing regulatory experience and decreasing the time and costs required for regulatory oversight, ultimately increasing the competitiveness of the advanced-technology firms (Lavoie and Sheldon, 1999). It is also argued that too much regulation is not optimal for technological innovation, because it confers advantage on large established firms who have the resources to deal with complex regulations (Zilberman *et al.*, 1997). This decreases the diffusion of technology and the number of potential commercial applications, which in turn decreases consumer choice. In other words, the social-rationality approach, opposed to multinational life-sciences firms, may actually be supporting the competitive position of these firms by calling for greater regulatory oversight to meet public fears, because only these firms have the resources necessary to deal with the complex regulations.

The social-rationality approach to regulating new technologies supports a reversal of the burden of proof, so that the focus of risk assessment is on minimizing the Type II risk (accepted when unsafe). In this case, GM crops and food produced from GM crops are assumed to be unsafe until they can be conclusively proved to be safe. In fact, some critics want the burden of proof not only reversed but also extended to include other legitimate factors as well. GM crops and GM foods would be rejected unless they could be proved safe and socially beneficial and hence meet all four regulatory hurdles as well.

It appears that the keys to this debate are the type of risk targeted and the principle of substantial equivalence. For instance, if the burden of proof were reversed and the critics were arguing from the perspective of speculative risks about non-equivalent GM crops, then GM crops would never be approved, because there would be no way to disprove the logical possibility of Type II speculative risks. Further, there would be no accepted analytical methods for identifying and then minimizing the speculative Type II risk. The implication is that regulatory responses would be biased against the techniques of modern biotechnologies and in favour of traditional plant breeding, even though traditionally bred crops are also associated with speculative risks. Yet reversing the burden of proof in risk assessment appears unnecessary if risk assessment is based on hypothetical risk and GM crops are substantially equivalent to the same-use conventional crops. In this case, focusing on hypothetical risks means that accepted analytical methods may be employed to identify and minimize Type I risks. Moreover, there would be no scientific or empirical reason to shift from the minimization of Type I risks to Type II risks, because GM crops would be considered substantially equivalent.

5.1.6 Risk tolerance in risk management

Once the risk assessment is complete, determining what level of risk is acceptable is a risk-management procedure. The standard approach is to establish tolerance levels, because zero risk is not possible: 'the concept of zero risk to the consumer from the food supply although attractive and the ultimate aim of food safety, given the current level of knowledge is not a feasible option' (Majewski, 1997). Despite this reality, there have been pressures for zero-tolerance risk-management policies associated with GM-crop regulations. The controversy arises because risk management must balance the economic interests of supporters, who support a risk-tolerance level for the presence of GM material in food products, with those social interests who demand zero risk (Table 5.4).

Tolerance levels are established for the presence of many types of possible materials in food products, such as the content of rat hair in macaroni, bugs in flour or pesticide levels on vegetables.[7] Organic labelling policies also accept tolerance levels. For example, the European Union (EU) organic labelling regulation (EC 2092/91) states that a food product must have at least 95% organic ingredients in order to be legitimately labelled as organic. Also, while some want zero-tolerance policies on the presence of GM materials in food products, a specific level of pesticide residue is tolerated. Pesticide residues and GM material are conceptually similar in the sense that they are generally in food products without the consumer knowing, or at least being sure. Yet there are important differences. First, GM crops have not been shown to have even hypothetical safety and health risks, while there is considerable information about the recognizable safety and health risks of pesticides. Secondly, while humans are designed to digest proteins, they are not designed to digest synthetic chemicals. Scientific-rationality supporters would argue that it is an irrational result that synthetic chemicals with known recognizable risks are tolerated to a certain level, while protein-based genetic modifications that have not been to shown to have even hypothetical risks to human safety and health are the target of zero-tolerance pressures.

Table 5.4. Risk tolerance in the risk-analysis framework.

Scientific rationality	Social rationality
Tolerance levels established	Zero tolerance pursued

[7]See the Food and Drug Administration of the US Department of Health and Human Services website on permissible food adulteration.

5.1.7 Science vs. other legitimate factors in risk management

According to the RAF, risk assessment must be free of normative considerations to the extent possible, while the risk-management process is where the rights and interests of all stakeholders must be balanced. There are three debates associated with this.

The first controversy arises because, as was shown in Chapter 4, many of the critics base their rejection of GM crops upon a belief system that is not open to compromise or concession. For instance, rejection of GM crops because of moral, ethical or religious beliefs about genetic engineering are not usually open to compromise.[8] Another example is critics of GM crops who support organic farming. Such critics often reject GM crops because of their rejection of the broader paradigm of intensive agricultural production. Instead of comparing an insecticidal GM variety of maize with a non-GM variety in terms of the benefits accruing because of the reduction in pesticide used in an intensive system, they compare the GM variety with an organic variety in an organic system and conclude that GM varieties are just not good enough. The problem here is that supporters and critics are arguing from different frames of reference and it is difficult to achieve compromise between different frames of reference in order to establish an economically and socially acceptable risk-management regulatory approach (van den Daele *et al.*, 1997).

The second debate arises because some argue that risk management should be based on scientific risk assessment and it should remain as objective and neutral as possible, not a venue for political compromise and concession. According to this view, the objective of risk management is the reduction and prevention of scientifically determined actual risk and this can be achieved without normative influence or consideration of perceived risks or other legitimate factors. For instance, proposed US regulations are analysed by the Office of Management and Budget to ensure that they are capable of achieving desired outcomes and that they are not the product of political objectives (see Chapter 6). According to this analysis, the risk-management strategy is expected to focus on the reduction and prevention of hypothetical risk, not on the assuaging of public fears and concerns about speculative risks.

The third debate arises because, if it is accepted that risk management is inherently a political function, then it would seem that it must be performed by elected officials. Yet it has been argued that consumers do not believe that elected officials can appropriately play this role (Grove-White *et al.*, 1997). If this is the case, then who should take the risk management decisions? The responsibility appears to slide back

[8]Ackerlof and Dickens (1982) argue that consumer beliefs persist over time and are difficult to change.

down to the scientific risk assessors (who must necessarily be free of normative factors in order to arrive at independent risk information) and yet the public allegedly does not trust the scientists either. How can regulatory development escape this conundrum? It has been argued that non-governmental organizations can fill this role by acting as honest brokers in precautionary regulatory development (Grove-White *et al.*, 1997). However, from the scientific-rationality perspective, this results in the entirely unacceptable situation of unaccountable third-party interest groups being responsible for regulations.

5.2 Principles of GM-crop Regulations

The seven general debates associated with the RAF reveal the existence of a significant range of views on the appropriate way to regulate advanced technologies. This gives rise to a challenging exercise of compromise and concession in the regulatory development process in an effort to balance technological progress with technological precaution. This challenge is exacerbated, however, by the development and commercialization of GM crops, because there are five specific debates over the fundamental principles for regulating them.

The objective of this section is to identify the five specific regulatory debates. The key point is that, while two jurisdictions may claim to have risk-analysis-based regulations, the range of views and debates means that the approaches can, in fact, be significantly divergent.

5.2.1 The precautionary principle

Most countries claim to base their GM-crop regulations on the precautionary principle. Yet the use of the precautionary principle in GM-crop regulations is associated with a significant degree of misunderstanding. In general, the precautionary principle implies that, in the face of uncertainty, regulators must overestimate risk.[9] However, in an operational sense, the precautionary principle has two somewhat different interpretations and regulatory jurisdictions can employ either of the two interpretations. The first interpretation, supported by scientific rationality, considers it to be a scientific risk-assessment tool while the second, supported by social rationality, considers it to be a social risk-management tool (Table 5.5).

As a scientific risk-assessment tool, the precautionary principle refers to the manner in which hypothetical risks are assessed. Recall

[9]For a summary of the precautionary principle, see Tickner (1997, 2001), Sanford (1998), Foster *et al.* (2000), Saunders (2000), Appell (2001) and Tickner and Raffensperger (2001).

Table 5.5. Precautionary principle in the risk-analysis framework.

Scientific rationality	Social rationality
Risk-assessment tool	Risk-management tool
Focus on hypothetical risks, with risk-averse likelihood functions	Focus on speculative risks, for which a risk assessment cannot be performed
Used in the WTO's SPS Agreement, the Codex Alimentarius and the IPPC	Used in the Cartagena Protocol on Biosafety

WTO, World Trade Organization; SPS, Sanitary and Phytosanitary Standards.

that hypothetical risks are those that lack experience or data, but, with the help of assumptions and/or likelihood functions, risks may be assessed within an accepted analytical method. Applying the precautionary principle at this stage would entail setting very risk-averse assumptions or parameters in the likelihood functions to ensure that uncertainty is sufficiently dealt with. Then hypothetical risks are calculated by credible experts, according to accepted analytical methods and according to the traditional burden of proof. The hypothetical risks are 'precautionary', because the precautionary principle has been built into the risk-assessment process. This approach accepts that regulators can only be reasonably certain of no adverse affects. It still supports the research, development and commercialization efforts that further the knowledge base and lead to greater understanding of the risks. Further, the hypothetical risk-assessment information is then used to advise the risk-management procedure.

As a social risk-management tool, the precautionary principle is the vehicle for allowing speculative risks into the risk-management procedure. Recall that speculative risks lack experience, data, a causal–consequence mechanism and an accepted analytical method for assessment. They are irrefutable logical possibilities, where a scientific risk assessment cannot be performed. Instead, speculative risks are 'best guesses' introduced into the risk-management procedure with no analytical or empirical method of evaluation. The social interpretation has its roots in environmental literature (Tait and Levidow, 1992). The Rio Declaration stated that:

> [i]n order to protect the environment, the precautionary approach shall be widely applied by states according to their capabilities. Where there are threats of serious or irreversible damage, lack of full scientific certainty shall not be used as a reason for postponing cost-effective measures to prevent environmental degradation.

The concern is that ecosystems are so poorly understood that precaution must be taken in approving the environmental release of products that may have irreversible or irreparable impacts. Anticipating the worst, regulators pursue zero risk in the sense that, under the precautionary principle, they must be certain of no adverse affects. They heavily scrutinize research and development and do not permit commercialization until it can be proved to be free of risk, essentially reversing the traditional burden of proof. Further, because the risk-management procedure involves balancing the rights and interests of supporters and critics, many different actors can introduce speculative risks into the risk-management procedure, which is unlike hypothetical risk assessment, where the risk information is compiled by credible 'experts'.

Clearly, the interpretation of the precautionary principle employed can lead to significantly different regulatory outcomes. The scientific interpretation is favoured by international standards-setting institutions, such as the Codex and the IPPC. It is also the interpretation under Article 5(7) of the Agreement on Sanitary and Phytosanitary Standards (SPS Agreement). The social interpretation has been established in the Cartagena Protocol on Biosafety and is supported by social-interest groups (Consumers' Association of Canada, 1994; Transatlantic Consumers Dialogue, 1999a).

The dual interpretation of the precautionary principle has created the unusual situation where Canada and the USA claim that Codex employs the precautionary principle, while the EU claims that it does not. In fact, the dual interpretation can be expected to cause trade tensions. Consider, for instance, that the IPPC employs the scientific interpretation in developing phytosanitary measures and yet the environmental tradition is to use the social interpretation. As members enact phytosanitary measures against GM crops, these measures will be evaluated in terms of their scientific justification. In this case, the Technical Barriers to Trade (TBT) Committee or the World Trade Organization (WTO) dispute-settlement mechanism will be charged with making the unenviable decision on which interpretation of the precautionary principle is acceptable and which is not.

5.2.2 Regulatory focus: products vs. processes

Domestic regulations on biotechnology focus either on the products created through the use of biotechnology or on the processes (i.e. how the products are made). This is the so-called products vs. processes debate.

Regulating products, not processes, is the traditional regulatory approach to regulating advanced technologies. Product-based regulations for GM crops and products derived from GM crops would be congruent with the traditional approach. This is the regulatory approach supported by scientific rationality and focused on application-based, novelty regulations and, of course, flows from framework regulatory principles, such as hypothetical risks and substantial equivalence.

On the other hand, regulating according to the use of modern biotechnology, especially genetic modification, is not congruent with the traditional approach. Modern biotechnology is employed across many sectors, such as agriculture, forestry, fisheries, pharmaceuticals, medicine and the environment. These sectors have been regulated independently, pursuant to specific, often divergent, mandates. Regulating the process requires regulations applicable to all sectoral applications, despite the divergent mandates. Social interests regard this approach as more capable of ensuring technological precaution (Table 5.6).

5.2.3 Regulatory structure: vertical vs. horizontal regulations

Closely associated with the products vs. processes debate is the decision to regulate biotechnology according to existing, vertical regulations or via new, horizontal legislation (Table 5.7). Regulating products of biotechnology is congruent with the existing, vertical regulatory jurisdictions in most countries. The key result is that existing vertical jurisdictions may be used to regulate GM crops, building on the accumulated expertise, capacity and trust in vertical agencies. This is the approach supported by scientific rationality, because it is viewed as the best way to ensure technological progress.

On the other hand, regulating the process usually requires the development of new horizontal regulations addressing issues that existing, vertical structures may not. Given that the fundamental features of biotechnology remain virtually the same regardless of the application, regulating the technology potentially provides a more integrated approach than using divergent regulations in different vertical agencies. As the mandates of vertical regulatory agencies were, in most cases, developed prior to the biotechnological revolution, it is argued that they were not designed to address the risks associated with this new technology (Rogers, 1990). A new, horizontal regulatory approach focused on the technology could possibly provide a more appropriate level of oversight, regardless of the product application. New horizontal regulations, however, face the inevitable political

Table 5.6. Regulatory focus in the risk-analysis framework.

Scientific rationality	Social rationality
Product focus	Process focus
Application-based	Technology-based
Employs the substantial-equivalence principle to determine appropriate regulatory oversight	Rejects the substantial-equivalence principle; all GM crops are novel

Table 5.7. Regulatory structure in the risk-analysis framework.

Scientific rationality	Social rationality
Vertical, existing structure	Horizontal, new structure
Regulatory authority retained by traditional agencies according to a product or application focus	Regulatory authority transferred to horizontal agencies underlying the traditional sectoral approach
Amendments to traditional legislation to deal with novel applications	New legislation to deal with process or technology

challenges of appeasing all actors, which is time-intensive and can lead to over-regulation and anticompetitive restrictions on the market (Cantley, 1998).

5.2.4 Regulatory decision-making in risk management

There are two aspects of regulatory decision-making to consider: (i) the type of decision-making procedure employed; and (ii) the participation in the decision-making procedure.

With respect to the type of decision-making, it can be either judicial or consensual. Judicial regulatory decision-making is done, for instance, through voting procedures among all the participants or among appointed 'judges'. After hearing the scientific evidence, a judgement is made. Consensual decision-making, on the other hand, requires the unanimous support of all participants before regulations may be enacted.

Similarly, in respect of participation in the regulatory decision-making procedure, it can be either the narrow participation of 'experts' or the broad participation of a wide range of actors. For instance, in a traditional, vertical, independent regulatory jurisdiction, such as agriculture, regulatory development and decision-making generally have a narrow set of involved actors, such as farmers, agrochemical companies and food industries. Issues associated with the regulation of GM crops have tended to dramatically increase the number of interested actors beyond this narrow set, because of the broad range of consumer concerns involved. The greater the number of actors, the greater will be the number of competing interests and, by implication, the regulatory decision-making procedure will be more complex.

In Table 5.8 there is a cross-comparison of regulatory decision-making. In general, judicial decision-making creates winners and losers; however, it circumvents the need to build unanimity. Judicial and narrow decision-making is the most timely and specific type of decision-making, although it may lack broader credibility, because it is exclusive and elitist. In fact, the discussion on the Codex revealed that Codex decision-making has shifted from a consensual to a judicial basis, because of the linking with trade agreements causing criticism about its credibility in establishing harmonized international regulations. In contrast, consensual and wide decision-making is the most credible, because it is inclusive; however, it is time-intensive and lacks specificity, with so many interests and concerns involved.

5.2.5 Regulatory instruments: mandatory labelling policies

Labelling is a risk-communication tool, along with public-service announcements, advertisements and public seminars, transferring knowledge from the supply side to the demand side of the market (Phillips and Isaac, 1998; Caswell, 1999). Labelling is often a voluntary private instrument that firms are both able and willing to pursue in order to differentiate their products in the market-place. Voluntary labels require a consumer willingness to search out and to pay for the differentiated products.

Labelling may also be a mandatory public instrument. In fact, mandatory labelling regulations for products made from GM crops are currently an important issue on the GM-crop regulatory agenda, because many states have developed or are developing mandatory schemes.

There are two motivations for mandatory labelling schemes: labelling for safety reasons or labelling for the consumer's right to know (Table 5.9). The former is the traditional approach, supported by scientific rationality, where the label is used to connote some kind of safety or health risk from the product, such as warning labels on cigarette packages or labelling the presence of allergens in a food product. According to the scientific-rationality perspective, GM crops determined to be substantially equivalent to non-GM crops can only face a mandatory labelling strategy if there is a safety or hazard justification.

Table 5.8. Regulatory risk-management decision-making in the risk-analysis framework.

	Narrow participation	Wide participation
Judicial/voting	Most timely Creates winners and losers Elitist/exclusive Employed in WTO	More time required Creates winners and losers More inclusive/credible Employed in EC Parliament
Consensual	More time required Limits winners and losers Elistist/exclusive Traditionally employed in Codex, IPPC	Most time-intensive Limits winners and losers Most inclusive/credible Employed in BSP

EC, European Community; BSP, Biosafety Protocol.

Table 5.9. Labelling policies in the risk-analysis framework.

Scientific rationality	Social rationality
Mandatory Public good Two reasons: 1. Substantially equivalent Safety or hazard reason for mandatory labelling Product or application focus 2. Novel Consumers' right to know reason for mandatory labelling Product or application focus	Mandatory Public good One reason: 1. Novelty of all GM crops (no substantial equivalence) Consumers' right to know Process or technology focus
Voluntary Private good Based on consumer willingness to pay	Voluntary Private good Based on consumer willingness to pay

However, GM crops determined to be novel can face a mandatory labelling strategy, in order to meet the consumers' right to know about novel products or applications of biotechnology.

With respect to the social-rationality position on mandatory labelling, as this paradigm rejects the substantial-equivalence principle and focuses on the process or technology used, all GM crops would be considered novel; and, similarly to the scientific-rationality approach above, in the event that all GM crops are considered novel they would face a mandatory labelling strategy, in order to meet the consumers' right to know that GM crops have been used in the processing or production of the food product.

There are both advantages and disadvantages to mandatory labelling schemes based on the consumers' right to know. It is advantageous in the sense that it meets a consumer demand to know whether the food product contains or has been derived from any GM material. It may provide for informed consumerism, because it is information that can build trust and ensure choice. However, labels not based on health and safety information are normative social statements, appeasing only those interests that have raised speculative, unsubstantiated claims. This provides undeserved credibility for the critics and creates the false impression that GM crops and foods are unsafe or inferior to conventional products.

The UK has adopted a mandatory labelling scheme based on the consumers' right to know, not based on safety considerations. This is a second-best policy, in the sense that it is a reaction to public fears, not to safety. The cost burden falls on the retail food outlets (supermarkets, caterers, restaurants, etc.), who are on the front line between producers and consumers. Therefore, this second-best labelling policy for the consumer's right to know has led to an inequitable distribution of costs. Further, given the credence nature of GM crops, there is a practical issue associated with labelling for the consumer's right to know; does a label indicating the presence or possible presence of GM material really allow for informed choice? While the label would allow consumers to know, it does not add to true informed choice, because, as shown in Chapter 2, overall consumers still remain uninformed about GM crops.

Recent research supports the importance of labelling to meet the consumer's right to know and it concludes that choice is paramount for consumers' risk perception. It also concludes, however, that, in the case of recombinant bovine somatotrophin (rbST) in milk, labelling brings premiums for the voluntary label on non-rbST products (Zepeda *et al.*, 1999). The implication is that a voluntary label for non-GM foods is both economically more efficient and consumer-beneficial than a mandatory labelling scheme not rooted in health or safety concerns. A voluntary labelling programme would more equitably distribute the costs of labelling according to consumers' willingness to pay for GM-free alternatives.

5.2.6 Summary of regulatory debates

Together, the seven debates associated with the regulatory framework and the five debates associated with the specific GM-crop regulatory principles reveal two distinct regulatory paradigms – a scientific-rationality paradigm and a social-rationality paradigm – where, essentially, these two paradigms establish the regulatory parameters of a jurisdiction (Table 5.10).

There are two crucial features of the GM-crop regulatory development process. First, and perhaps the most important feature, is the fact that the fundamental regulatory principles have not been established. For instance, the definition of 'risk' for regulating new technologies and the applicability of the substantial-equivalence principle in regulations remain highly contentious debates. As a result, GM-crop regulations are unstable, as various economic and social interests lobby for either the scientific- or the social-rationality approach. Moreover, the regulatory instability associated with GM crops at national and international levels is a unique feature, distinct from other social regulatory barrier issues. For instance, while the transatlantic trade of beef products has been controversial on health and safety grounds, at least the fundamental regulatory principles for regulating beef safety are very much universal (Spriggs and Isaac, 2001). Until the fundamental

Table 5.10. Comparison of scientific- and social-rationality regulatory approaches.

	Scientific rationality	Social rationality
General or framework regulatory debates		
Belief	Technological progress Enhances growth and development leading to higher social regulations – a regulatory race to the top	Technological precaution Technology cannot be separated from socio-economic factors – must be socially responsive
Type of risk	Recognized Hypothetical	Recognized Hypothetical and speculative
Substantial equivalence	Yes	No
Science or other factors in risk assessment	Safety Health	All four regulatory hurdles Safety Health Quality Other legitimate factors
Burden of proof	Innocent until proved guilty Minimize Type I (rejected when safe)	Guilty until proved innocent Minimize Type II (accepted when unsafe)
Risk tolerance	Minimum	Zero tolerance
Science or other factors in risk management	Safety or hazard-based Risk management is for risk reduction and prevention only	Broader socio-economic concerns Risk management is for social responsiveness
Specific regulatory debates		
Precautionary principle	Scientific interpretation Risk-assessment tool	Social interpretation Risk-management tool
Focus	Product-based, novel applications	Process- or technology-based
Structure	Vertical, existing structures	Horizontal, new structures
Participation	Narrow, technical experts Judicial decision-making	Wide, 'social dimensions' Consensual decision-making
Mandatory labelling	Two scenarios Substantially equivalent Safety or Hazard basis Novel Consumers' right-to-know basis	All GM crops are novel Consumers' right-to-know basis

principles for regulating GM crops are established, regulatory instability will remain and regulations will shift within the regulatory parameters according to the influence of the competing interests.

Another crucial feature of the GM-crop regulatory development process is the path dependency of the regulatory paradigms. Initially, the particular regulatory trajectory is determined by the independence and discretion of the regulators and the role of science in the regulatory development process. If scientific rationality is the tradition, then subsequent regulatory principles tend to hold an important role for science in determining the appropriate level of regulatory oversight. In fact, the position on the various regulatory debates cascades from the paradigm followed. For instance, the scientific-rationality paradigm leads to a focus on hypothetical risks and an acceptance of the substantial-equivalence principle for determining the novelty of product applications, which in turn lead to the use of the precautionary principle as a risk-assessment tool, a product-based, vertical regulatory structure employing narrow and judicial decision-making and supporting mandatory labelling in the instance of hazard or safety only. Conversely, if social rationality is the main paradigm, then the underlying support for technological precaution encourages a process- or technology-based focus, which cascades into the use of the precautionary principle as a risk-management tool and into a process-based, horizontal regulatory structure with wide and consensual decision-making and mandatory labelling for the consumers' right to know in order to be socially responsive.

Additionally, the path dependency is also influenced by the competitive position of the jurisdiction in the new technology. A commercialization lead affords a jurisdiction more time to become acquainted with the credence factors of a new technology and may even allow the unknown risks to become known. On the other hand, a commercialization lag ill prepares the consumers within a jurisdiction for the sudden appearance of advanced technologies developed elsewhere.

The path dependency of the two regulatory paradigms makes integration very difficult. For instance, a regulatory trajectory focused on technology-based oversight is difficult to combine with a trajectory focused on application- or product-based legislation according to novelty. Similarly, substantial equivalence can be either accepted or rejected within a jurisdiction; there is really no middle ground with this regulatory principle. Further, a jurisdiction that follows the scientific-rationality paradigm would find it difficult, in the end, to implement a mandatory labelling strategy that is completely process-based for the consumers' right to know. Indeed, like the labelling policy, many of the regulatory positions in the two paradigms are not stand-alone positions but are part of a logical flow of regulatory decisions that can be traced all the way back to the fundamental view on the approach to technology in society: progressive or precautionary.

The two distinct paradigms, along with the regulatory debates identified in Table 5.10, can be used to characterize the GM-crop regulatory development process within a regulatory jurisdiction. In order to do so, the commercial position of the jurisdiction, the traditional regulatory role of the state and the influence of the various economic and social interests must be identified. This will be done for the North American and the EU regulatory approaches in Chapters 6 and 7, respectively. However, prior to this, it is now vital to link the regulatory approaches in Table 5.10 with various regulatory integration strategies. This is the objective of the next section.

5.3 Regulatory Integration

While two jurisdictions may claim to have risk-analysis-based GM-crop regulations employing the precautionary principle, the previous sections have revealed that their approaches can, in fact, be drastically different. The result is the creation of social regulatory barriers and regulatory regionalism, as GM crops approved for market access in one jurisdiction face regulatory delay and perhaps rejection under the regulatory approach within another jurisdiction.

As discussed in Chapter 1, social regulatory barriers represent a trade challenge quite unlike traditional border measures, such as tariffs and quotas, which tend to be quantitative, easily identifiable and driven by commercial protectionism. Instead, social regulatory barriers are non-quantitative and involve social preferences and concerns. Regulations for food safety and environmental protection tend to be inwardly focused, reacting only to domestic interests, with little tradition for considering their potential market-access impacts. Although regulatory barriers to trade are at present a limited source of market-access barriers,[10] their use in the trade of GM crops is already a significant transatlantic trade tension.

In this section, the prospects for the integration of GM-crop regulations, given the possible divergence in approaches between jurisdictions, are examined. Essentially, regulatory integration is a function of the regulatory approach adopted. A scientific-rationality approach to regulating GM crops tends to be congruent with an international economic-integration strategy supported by the international trade regime. Alternatively, a social-rationality approach tends to be congruent with an international social-integration strategy incongruent with the traditional trade approach.

Therefore, social regulatory barriers incur a conflict between economic interests, which support scientific rationality and shallow economic integration and social interests, which support social rationality and deeper social integration. This regulatory integration conflict will be illustrated with reference to several trade disputes involving issues that may be reasonably extrapolated to disputes involving regulatory barriers

[10]Roberts (1998) concludes from a comprehensive evaluation of market-access barriers facing US agricultural exports that most barriers are imposed to protect the economic interests of domestic producers, not to protect the concerns of domestic consumers.

to GM crops. Then, based on this analysis, the three choice parameters for regulatory integration identified in Chapter 1 will be assessed.

5.3.1 Analysis of trade agreements and GM-crop measures

In order to illustrate the difference between economic- and social-integration strategies, two trade-dispute cases and one current trade tension are summarized. To control the development and commercialization of GM crops, some members of the WTO have proposed regulatory measures, such as an outright ban on GM crops and GM foods or mandatory labelling schemes for GM foods. Both food-safety and environmental-protection justifications have been cited to support these measures. Are these measures compatible with the rights and obligations of members under the WTO? These examples will be examined in order to understand how food-safety- and environmental-protection-type social regulatory barriers restricting market access for GM crops and GM foods may be dealt with according to the economic-integration approach. This assessment reveals the significant limitations of the WTO approach when dealing with social regulatory barriers. The implications of this for trade-diplomacy efforts will be assessed following the case-studies.

A. Food-safety ban: EU and Canada/US beef-hormones dispute

The use of hormones in the production of beef has been a controversial issue in Europe for over 20 years. In March 1988, an EU Directive (88/146/EEC) banned the use of six hormones in beef production in both domestic and foreign (imported) beef products. The European Commission (EC) directive identified six separate growth hormones and applied a zero-tolerance policy.[11] As a result, North American beef containing such hormones was effectively prohibited from EC markets, regardless of the fact that hormone use had been approved as safe in North America.

In 1988, the ban was not in violation of the General Agreement on Tariffs and Trade (GATT) 1948, because it was being applied according to the principles of non-discrimination – that is, it was being applied to all beef producers, both domestic and foreign, and equally to all like or substantially equivalent products. Yet, in 1995, when the WTO came into force, those Parties of Export who felt that the ban was scientifically unjustified protectionism now had the legitimate opportunity to challenge the ban under the safety-related provisions of the SPS Agreement.

[11] Three naturally occurring ones (oestradiol-17β, progesterone and testosterone) and three synthetic ones (trenbolone acetate, zeranol and melengestrol acetate (MGA)).

Canada and the USA complained, on behalf of their domestic beef producers, to the WTO that the EU regulations against imports into Europe of beef treated with growth hormones violated the EU's obligations under the SPS Agreement.[12] They argued that the ban was not consistent with Codex standards and that there was insufficient scientific justification for the EU to impose a standard higher than the Codex standard. The Codex ruled, in 1995, on the maximum residue limits for the safe use of these hormones in beef production (CAC, 1995; see also McDonald, 1998). Also in 1995, an EU conference reported that the hormones were safe and within normal physiological ranges (*Europe Drug and Device Letter*, 1996). Yet, despite the Codex standard and even the advice from its own scientists, the European Commission maintained the ban.

The WTO dispute-panel decision ruled against the EU ban, and the EU promptly appealed the decision. In its decision, the Appellate Body of the WTO was careful to argue that its decision was a 'procedural ruling' on the basis of the wording of the SPS Agreement and not a decision about the permissible use of health and safety regulations by members. In fact, it has been argued that the Appellate Body does make efforts to uphold the sovereignty rights of members when making decisions about their obligations under the trade agreements (Jackson, 1996).

The Appellate Body ultimately held that the EC directive did violate the SPS Agreement, because the measures were not based on an appropriate risk assessment and were not scientifically justifiable. No risk assessment had been done for the melengestrol acetate (MGA) hormone and scientific evidence from both North America and Europe along with the Codex standard, suggested that the other five hormones were safe if properly administered. The evidence presented concerning the carcinogenic potential of increased hormone ingestion was found to be unrelated to the six hormones in question. The panel also found no evidence that any of the six hormones were being improperly administered on a widespread basis.

Further, the Appellate Body also considered the justification for the ban under SPS Article 5(7), the use of the precautionary principle. There was found to be an insufficient justification because: (i) a provisional ban must be temporary while the ban was a permanent EU directive; and (ii) the ban is justified if there is insufficient scientific evidence and yet there were numerous scientific studies from both Europe and North America. On this matter, the Appellate Body concluded that 'the precautionary principle has not been written into the SPS Agreement as a grounds for justifying SPS measures that are otherwise inconsistent with the obligations of Members set out in particular

[12]European Communities – Measures Concerning Meat and Meat Products (Hormones); Panel, 18 August 1997.

provisions of that Agreement' (Appellate Body Ruling Article 124). This ruling may be interpreted as requiring members to employ the scientific interpretation of the precautionary principle – that is, members must prove more than just public fear to support a claim of precaution; only hypothetical risk assessments or highly conflictive scientific conclusions are an appropriate justification. In the hormones case, even the EU scientists found no evidence to justify the ban. The Appellate Body of the WTO therefore ruled against the EU ban and the EU's use of the social interpretation of the precautionary principle, recommending that the EU bring its measures into conformity with its SPS obligations.

This order has yet to be implemented. The EU informed the WTO that it must complete a new risk assessment and then implement new SPS measures to continue to ban hormone-treated beef. An arbitrator had ordered that the EU must comply with the panel ruling within 15 months, which was by 18 May 1999. This should serve as an important warning; the EU is willing to remain in violation of a WTO dispute-settlement decision in order to uphold its social regulations.

The WTO Appellate Body ruling raises the question as to what is a sufficient scientific risk assessment under the WTO. With respect to the regulatory debates assessed above, the answer is that a legitimate risk assessment must be a comprehensive hypothetical risk assessment focused on the identifiable hazards to support the trade-restricting measure. Risk assessments based on vague, non-specific speculative risks or the social interpretation of the precautionary principle would not be deemed legitimate assessments according to these principles.

An important point is that the economic-integration approach of the WTO is directly in conflict with the social-integration approach supported by the European Court of Justice (ECJ). For instance, the ECJ decided in favour of the ban, because it was willing to consider the validity of non-science and non-safety objectives. The ECJ ruling claimed that 'traders were not entitled to expect that a prohibition on administering the substances under question to animals could be based on scientific data alone' (Scott, 1998). This difference should be expected because, while the WTO is pursuing the traditional trade model of shallow, economic integration, the role of the ECJ is to facilitate the deeper social integration of EU member states. In this sense, the ECJ should be considering broader socio-economic issues, while such issues are arguably not in the remit of the WTO.

By extension, any EU ban on GM foods must fulfil the same scientific justification and risk-assessment requirements in order to be trade-compliant. Scientists in North America and in Europe have consistently argued that there are no justifiable grounds for banning GM foods on the basis of food-safety concerns, other than the conventional concerns about allergens, which are associated with non-GM foods as well. In fact, in most cases the unilateral bans by EU member states on particular GM crops have resulted from politicians overruling the advice of their own scientific advisory committees, because of consumer concern (see Chapter 7). Therefore, at the present time, an EU ban on GM foods, either permanent or provisional, appears to violate the WTO's SPS Agreement, because there is not a sufficient scientific justification to declare that GM foods are unsafe.

Such a WTO decision on social regulatory barriers would surely be condemned for three reasons. First, it adopts a rigid scientific-rationality approach to the RAF (Table 5.10), requiring a focus on hypothetical, not speculative, risks. Secondly, it limits the social responsiveness of risk management, because it requires the evidence of actual risk to impose a ban, not simply perceived risk. Thirdly, and perhaps most crucially, it supports the scientific interpretation of the precautionary principle as a risk-management tool over the social interpretation favoured by social interests.

B. Environmental-protection ban: tuna–dolphin and shrimp–turtle

The tuna–dolphin trade dispute was based on a Mexican challenge to a US import ban on tuna products where so-called dolphin-friendly catch techniques had not been used. The USA cited evidence about the recognized risk of certain tuna catch techniques on dolphin mortality rates and concluded that certain techniques were unacceptably risky. Consequently, tuna products in which the risky techniques were used faced an import ban. The dispute was pre-WTO and it went before the dispute settlement mechanism of the GATT.

Essentially, the US import ban was an environmental-protection measure taken against the non-safety-, non-product-related PPMs of imported products. The method of catch had no impact at all upon the safety or quality of the tuna product, instead it had an environmental impact upon dolphins. The tuna imports were targeted with a second-best measure, an import ban designed to send a signal that US consumers prefer dolphin-friendly catch techniques.

The GATT dispute panel found in favour of Mexico. The ruling, congruent with the scientific-rationality paradigm, concluded that the USA had violated its GATT obligations by imposing a discriminatory import ban according to non-safety-, non-product-related PPMs. The ruling was sympathetic with US environmental preferences, however, and suggested that a voluntary 'dolphin-friendly' labelling scheme was the first-best measure to signal to consumers which non-product-related PPM had been used in the tuna catch. Such a scheme met two important criteria: first, it was a first-best and market-based strategy and, secondly, it avoided the use of unjustifiable trade barriers based on non-product-related PPMs and the extraterritorial extension of environmental preferences.

Similarly, the shrimp–turtle trade dispute was based on a challenge brought to the WTO by India, Malaysia, Pakistan and Thailand, to a US import ban on shrimp that were caught in nets that also caught and killed sea turtles. Similarly to the tuna–dolphin case, the shrimp–turtle import ban was based on the non-safety-, non-product-related PPMs of a traded product. The Dispute Settlement Body (DSB) panel ruled against the US import ban, which the US appealed. The WTO Appellate Body ruled that the US import ban violated the TBT Agreement by imposing a discriminatory trade ban according to non-product-related PPMs. Again, the alternative suggestion was a voluntary labelling scheme. Additionally, the Appellate Body suggested that the involved countries could negotiate a multilateral environmental agreement (MEA) based primarily on the environmental concerns, outside the WTO framework.

Both cases reveal that environmental trade measures based on preferences for non-safety-, non-product-related PPMs are not consistent with the rights and obligations of the international trade agreements. According to the WTO, a member cannot unilaterally impose second-best extraterritorial environmental preferences for non-product-related PPMs upon other members, because this violates the non-discrimination principle and the product focus of trade agreements. Instead, two alternatives are more effective: a voluntary labelling scheme connoting that a product has used an environmentally acceptable non-product-related PPM; and an MEA pertaining explicitly to the environmental concern at issue as a first-best measure to address the environmental problem.

By extension, these cases suggest that an outright import ban on GM crops based on non-safety-related environmental-protection justifications is not a trade-compliant strategy. First of all, there remains significant uncertainty about whether transgenic modifications to crops are even PPMs (see Chapter 3). Secondly, supposing that transgenic modifications were considered to be PPMs, the approved GM varieties have been approved on the basis of substantial equivalence; hence, the PPMs would be non-product-related. According to both the tuna–dolphin and the shrimp–turtle trade-dispute rulings, an environmental-protection ban based on non-product-related PPMs is not trade-compliant. The alternative, then, is a voluntary labelling scheme to connote the use of a non-product-related PPM, which is examined below, or an MEA, such as the Cartagena Biosafety Protocol (see Chapter 4).

Caution must be taken, however, in extending the tuna–dolphin and shrimp–turtle cases to GM crops. The reason for this is because of the unresolved controversy surrounding the social interpretation of the precautionary principle in environmental-protection objectives. In the cases of tuna–dolphin and shrimp–turtle, the recognized risk of either dolphin- or turtle-unfriendly techniques to their mortality was known. With GM crops, many of the environmental concerns used to support an outright ban are based on speculative risks, not recognized or hypothetical risks. Unlike food-safety issues, non-safety-related environmental-protection issues traditionally and legitimately use the social interpretation of the precautionary principle in risk management. Therefore, while food-safety justifications must employ hypothetical risk assessments and the scientific interpretation of the precautionary principle, non-safety-related environmental-protection justifications in reaction to speculative risks – some of which may be of dread consequence – may employ the social interpretation of the precautionary principle, permitting a much broader justification for a trade ban in order to protect environmental biodiversity.

In addition, a safety-related environmental-protection ban against GM crops, justified as protecting the safety of plants, natural fauna, etc., would fall under the jurisdiction of the SPS Agreement. In such a case, the regulations would have to meet the scientific-justification requirements established by the IPPC. Similarly to a food-safety ban, an environmental plant-safety-type ban on GM crops must be based on hypothetical risks, focused on safety only and with no scope for the use of the social interpretation of the precautionary principle. To date, there is insufficient scientific evidence that GM crops pose an environmental plant-safety risk necessary to justify such a ban.

Therefore, an environmental-protection ban on GM crops, either safety- or non-safety-related, would probably violate trade agreements. Yet such a decision would be widely condemned by social interests. In this case, the lack of legitimacy given to domestic environmental preferences (non-product-related PPMs) under the traditional trade framework would be an important source of dissatisfaction, because it fails to account for normative preferences in the regulatory measures. The analysis of the trade approach to an environmental ban also highlights the different interpretations of the precautionary principle. Again, the economic-integration approach essentially rejects the social interpretation.

C. Labelling GM foods

The labelling of products of modern biotechnology is an important regulatory integration issue currently facing Codex. Mandatory labelling policies are being developed by member governments without the international guidance of Codex, which has yet to establish a Codex standard on this issue (see Chapter 3). The purpose of this section is to examine the relationship between such a labelling policy and the Codex, the SPS Agreement and the TBT Agreement; is such a labelling policy trade-compliant?

There are two important aspects of mandatory labelling to consider. Labelling for food-safety reasons, such as for the presence of allergens, is an SPS issue. Labelling for non-food-safety reasons, such as the consumers' right to know, is a TBT issue.

With respect to mandatory labelling for food-safety reasons, such a policy would require a scientific justification for declaring foods produced from GM ingredients to be unsafe, similar to the food-safety ban discussed in A, above. If there exists evidence that foods produced from GM crops are unsafe, however, then the first-best measure would be a ban, not a labelling strategy. The discussion in regard to a ban on foods produced from GM ingredients is relevant here. Again, as there seems to be insufficient scientific justification for a safety-related ban, a mandatory labelling measure based on the safety of GM foods would not be in compliance with the SPS Agreement.

With respect to mandatory labelling for the consumers' right to know, a label would indicate that the product is made of GM material or was exposed to GM material during the production process. If there is no evidence that such food products are unsafe, but consumers remained concerned, then the first-best policy is to label the products, based on the consumers' right to know. The controversy arises because – as discussed previously – the consumer's right to know is not a universal justification for mandatory labelling. Consider two cases where the consumers' right to know was not used to justify mandatory labelling: the approved GM ingredient, rennet,[13] widely used in cheese production in Europe, and the GM bakers' yeast, approved for use in the UK in 1990. In both cases, the GM version was considered to be substantially equivalent to the conventional version, and there were no labelling requirements justified by the consumers' right to know about the use of either GM rennet or GM yeast.

There is, however, precedence for mandatory labelling based on the consumers' right to know about a food's PPMs: the Codex guidelines on labelling the use of food irradiation. Yet this precedent is not very relevant to the case of GM crops, for several reasons. First and foremost, in the case of irradiated meat, there is a Codex standard, first agreed in 1983 and amended in 1989. The issue of labelling GM foods at Codex has not produced a Codex guideline that supports the consumer's right to know. In fact, the only Codex recommendation on the issue, from the Codex Secretariat, supports mandatory labelling only for safety issues associated with novel GM products, not for the consumer's right to know about PPMs for non-safety reasons. Given this

[13]Rennet is a natural enzyme found in the stomach of veal calves and used in cheese making. A GM microorganism with a duplicated calf gene codes for the enzyme, which can then be industrially produced via fermentation rather than harvested from the gut of slaughtered calves (Harlander, 1993).

important role of Codex, it is useful to briefly discuss the development of this Codex recommendation.

The Codex Committee on Food Labelling (CCFL), currently chaired by Canada, considers international food-labelling issues; drafts labelling provisions (and amendments) that are applicable to all foods; and endorses labelling provisions in the standards, codes of practice, guidelines and recommendations prepared by other Codex committees. The CCFL has discussed the labelling of biotechnology products at five separate meetings without reaching a conclusive Codex position. At the first meeting (October 1994), the key issue was whether to develop a mandatory and comprehensive label scheme to apply to all foods derived from biotechnology or to apply only to novel products. On one hand, Australia, Canada and the USA argued that the science- and safety-based Codex was not the proper venue for consideration of the consumers' right to know product information about non-safety- and non-product-related PPMs. On the other hand, the EU position, supported by many civil-society organizations, was that the consumers' right to know must be a crucial objective of food standards. In 1995, at the request of the CCFL, the USA produced a position paper on biotechnology labelling, which argued for a case-by-case, product-based approach to labelling, not a broad mandatory policy on labelling the use of modern biotechnology in food production (Horton, 1997). At the 43rd Session of the Codex Executive Committee, the issue of mandatory labelling for other than safety reasons was also considered, but with no resolution (Codex Executive Committee, 1996).

At the second meeting (May 1996), the two general positions were reiterated. The CCFL requested that the Codex Secretariat prepare a discussion draft on the biotechnology-labelling issue for the next meeting.

At the third meeting (October 1997), the Codex Secretariat's discussion draft was presented. The two crucial recommendations from the Codex Secretariat were that mandatory labelling should: (i) only cover non-equivalent or novel products; and (ii) focus on health risks, including allergens. These recommendations were very consistent with the scientific-rationality position on mandatory labelling and they did not support the consumers' right to know about non-safety, non-product-related PPMs as a justification for a mandatory labelling policy. But other delegations could not agree to these recommendations. Little progress was made at the fourth meeting in May 1998. Some leading agricultural exporters, such as Australia, Brazil, Canada, New Zealand, Peru and the USA supported the adoption of both proposals, but many European delegations, along with India, continued to block agreement.

At the April 1999 CCFL meetings, there was still no success in establishing a Codex standard on a mandatory labelling policy for foods produced through modern biotechnology techniques. The US delegation, along with major agricultural exporters, reasserted support for the Codex Secretariat's recommendations. The German delegation, on behalf of the EU, supported a mandatory comprehensive labelling policy based on the use of biotechnology. Many delegations informed the Codex Committee that they were unilaterally developing mandatory labelling policies for the consumers' right to know, with or without the endorsement of Codex. They argued that essential or substantial equivalence was a useless term when the justification was the consumers' right to know. Consumers' International (CI), argued that the consumers' right to know must be the basis for the Codex labelling policy (CI, 1999a), as it was with irradiated meat. CI also supported the alteration of terminology to focus on GM, not on the use of modern biotechnology in general (CI, 1999b). The Canadian hosts of the CCFL proposed that the ambiguity around the use of the term substantial or essential equivalence should be clarified by a working group (CAC, 1999).

The regulatory integration issues associated with labelling GM foods is also on the agenda of the TBT Committee. Technical labelling requirements are justified under the TBT Agreement according to the protection of consumer health and safety and according to the consumers' right to know in order to prevent deceptive practices. As previously discussed, the USA submitted a request to the EU and the WTO's TBT Committee that the regulation (EU Regulation 1139/98 Soya/Maize regulations) be amended to reflect the trade concerns of the USA and other agricultural exporters, such as Australia, Canada and New Zealand. The submission claimed: (i) that the EU regulation did not achieve a legitimate objective; and (ii) that the implementation of the regulation was problematic and created an unjustifiable barrier to trade. With respect to the first claim, the USA argued that GM crops do not differ as a class from conventional varieties. As other GM techniques, besides transgenic modification, such as mutagenesis and somoclonal variation, do not have to be labelled as such, there is no justification for differentiating transgenic-crop varieties. With respect to the second claim, the USA argued that the regulation was not non-discriminatory, as required under the TBT Agreement, as it would discriminate between those exporters where GM crops are produced and exporters where GM crops are not produced. Further, in respect of the 1% tolerance threshold for adventitious contamination agreed in the EU, the USA was concerned that there was a lack of both standardized and accurate testing methodologies, such that different tests would produce different test results. In the event that testing methods between jurisdictions differed, this raised concerns about the liability of a positive test and, perhaps, a rejection of an export shipment.

Therefore, although not consensual, the current Codex Secretariat recommendation for the mandatory labelling of biotechnology-based products supports labelling for product-related PPMs resulting in novel crops only, not for the consumers' right to know about non-safety-, non-product-related PPMs. Recall that all Codex standards, codes, guidelines and recommendations are considered 'standards' according to both the SPS and TBT Agreements. As a result, the establishment of a mandatory comprehensive and extensive labelling scheme is sure to initiate a trade challenge and, in all likelihood, would be found non-compliant with the WTO for two reasons. First, there would be insufficient scientific justification for a mandatory labelling policy based on food-safety concerns under the SPS Agreement. Secondly, the labelling recommendation of the Codex Secretariat does not support the use of a mandatory comprehensive labelling policy for the consumers' right to know non-safety issues. Additionally, although a voluntary labelling scheme for products made from GM crops would be compliant with trade rules, a voluntary scheme is unlikely to be acceptable to social interests, because it would lack the sanctions to ensure that the consumers' right to know is met.

D. Discussion: implications for trade diplomacy

It appears that both an outright ban on the use of GM products, on either food-safety or environmental-protection grounds, and a mandatory and comprehensive labelling scheme would contravene international trade law. Yet a WTO decision against either a ban or a comprehensive labelling scheme would essentially be a ruling against the social-rationality approach to regulations, including the social interpretation of the precautionary principle and the right to consider other legitimate factors, such as consumer concerns, which may not be grounded in scientific evidence.

Given the distinct conflict between economic and social interests, the objective is to assess the implications for traditional trade diplomacy. Can the WTO approach facilitate the integration of food-safety and environmental-protection regulations for GM crops in order to prevent social regulatory barriers to trade? While there are strengths to the WTO approach, there remain crucial weaknesses, which seem to indicate that the economic-integration approach of the WTO cannot, and indeed should not, facilitate social regulatory integration. In other words, trade diplomacy must be amended if it is to remain a viable force in the international integration of GM-crop regulations.

On one hand, there appear to be several strengths of the WTO in facilitating regulatory integration. In general, the WTO is institutionally more capable of facilitating the centralization of domestic regulatory policies in order to encourage trade liberalization than was the international trade regime under the GATT 1948. There are three aspects to note. First, membership in the WTO is an important national priority, where members take the objective of trade liberalization and the ensuing rights and obligations of membership very seriously. Indeed, gaining membership is an important priority for non-member states (e.g. Russia and China). Secondly, the dispute-settlement mechanism is better equipped to deal more equitably with the market-access disputes that contravene the rights and obligations of WTO members (Jackson, 1996). In fact, since many of the disputes have involved either the EU or the USA (or both), the panels are often composed of smaller-country representatives, who have exercised important influence over DSB precedence so far. Also, as the dispute-settlement decisions are binding, despite their power, the USA and the EU are compelled to fulfil the obligations of a panel decision rather than employing the 'might is right' principle. Consider both the tuna–dolphin and the shrimp–turtle disputes. The USA lost both disputes to developing countries. In this sense, under the DSB, the obligations of the WTO are a gift from the strong members to the weak members. Thirdly, the Agreements on Agriculture, TBT and SPS measures have clarified the permissible role of national governments in establishing domestic regulations with potential trade impacts.

With respect to food safety specifically, the SPS Agreement is a rules-based agreement focused on hypothetical risk assessment and employing the scientific interpretation of the precautionary principle. Accordingly, certainty and stability are built into the trade rules and there is considerable scope for members to unilaterally impose social regulatory barriers, provided they are scientifically justifiable.

With respect to environmental-protection measures, the determination of members to honour their obligations possibly prevents members from making environmental commitments that they will not keep, which has been a familiar complaint about multilateral environmental agreements. The WTO has considerable experience with multilateral agenda setting among often very divergent national interests. This experience could be channelled towards establishing an international environmental agenda. The primary benefit of this approach is that an internationally consistent, rules-based regime may be developed. This is a desirable outcome in the sense that it would lend real international discipline and certainty to very contested environmental-protection issues and it would ensure that multilateral environmental externalities are dealt with. Additionally, any initiatives to come out of this framework would increase trade compatibility, eliminating or reducing both market fragmentation and trade threats.

Yet, despite the apparent strengths, there are four general weaknesses with the WTO approach. First, and principally, it is an institution focused on trade liberalization and based on the traditional trade model of economic integration and regulatory competition. Regulatory competition does not deal effectively with broader, populist issues, as the struggles of the Committee on Trade and the Environment (CTE) indicate. Instead, the WTO's regulatory-competition approach continues to encourage traditional, closed-door and non-transparent trade diplomacy. In fact, despite a few high-profile dispute-resolution cases, most trade tensions are dealt with according to the traditional trade-diplomacy model. The trade framework inappropriately deals with other legitimate objectives, for regulatory development and regulatory barriers to trade will be evaluated according to trade principles – their sufficient scientific evidence and whether or not they minimize trade disruption – and not according to whether or not they achieve their legitimate social objectives. In short, regulations based on non-scientific justifications are unlikely to be found trade-compliant, even though they may be socially responsive.

Secondly, when disputes are taken to the WTO, DSB decisions are 'forced' on members, creating winners and losers, increasing public dissatisfaction among losers and subsequently decreasing overall support for trade-liberalization efforts. In this context, it is difficult for members to reject calls for unilateral protectionism, since the trade agreements fail to accept socially responsive regulations.

Thirdly, the scientific basis for justifiable regulations and the scientific interpretation of the precautionary principle limit the scope for domestic regulators to respond to public concern. During the risk-management process, economic trade analysis often concludes that 'science' will be the final arbiter of trade disputes (see Buckingham *et al.*, 1999; Jackson, 1999). Yet there is no agreement on the legitimate and credible role of science. A common argument against the scientific basis of WTO agreements and the DSB decisions is that they are crucially flawed. They rely on international organizations, such as the Codex, to develop international standards. The Codex, however, can no longer be interpreted as establishing consensual international standards, because it has adopted judicial (voting) decision-making, which creates winners and losers and does not reflect international consensus. The focus on science has created a 'your scientist' vs. 'my scientist' ethos, where trade lawyers will define appropriate science, food-safety and environmental-protection regulations.

Fourthly, with GM crops, trade protection is demanded by social interests, while trade agreements have historically developed to deal with protectionist pressure from producers (Perdikis *et al.*, 1999). The question then arises: are some regulatory issues emotive enough for members, or a group of members, to tolerate non-compliance with the WTO rules? Such an outcome would have disastrous results for the institutional credibility and authority of the international trading regime. In fact, this has already happened with the EU–US beef-hormone dispute, where the EU is prepared to remain in permanent contravention of the WTO rules.

Apart from the general weaknesses, there are specific weaknesses in the WTO approach to dealing with food safety according to the scientific-rationality approach. EU regulators, along with some of its member states, have continually ignored the advice of its own scientists over beef-hormone, antibiotic and GM-crop approvals, in an effort to be seen as socially responsive. The WTO's scientific-rationality approach does not accommodate the social rationales, such as the social interpretation of the precautionary principle in risk management. A WTO decision against either a ban or a comprehensive labelling scheme would essentially be a ruling against: (i) the social interpretation of the precautionary principle; (ii) the consumers' right to know about non-product-related PPMs; and (iii) the right to consider other legitimate factors, such as consumer concerns, which may not be grounded in scientific evidence. As these are considered to be 'unalienable rights' within the social-rationality approach, such a decision would lead to condemnation of the WTO from social interests.

Similarly, in respect of environmental-protection measures, there are several specific weaknesses of the WTO. First, trade objectives may not be consistent with environmental objectives and, in the event that the objectives are not congruent, trade objectives will dominate. Secondly, the WTO lacks credible environmental expertise or the legitimacy to adjudicate environmental disputes and the CTE has struggled to develop a rigorous environmental agenda (*Economist*, 1999b). Thirdly, as a multilateral institution focused on 'top-down' approaches to enhancing trade, the WTO may be insensitive to legitimate national differences in environmental concerns, which are the basis for national environmental-protection measures. Fourthly, developing a 'top-down' environmental agenda acceptable to all members at varying levels of development risks the establishment of an agenda that reflects the lowest-common-denominator approach rather than the establishment of an agenda that reflects the highest environmental standards, an outcome completely unacceptable to many civil-society organizations. Fifthly, the science-based approach to phytosanitary measures under the IPPC is in conflict with the traditional environmental approach employing the social interpretation of the precautionary principle. Sixthly, the WTO provides inadequate participation in decision-making for social interests, such as consumer and environmental organizations. The members of the WTO are governments and it is expected that consumer and environmental organizations state their positions to the national delegations, not directly to the WTO. Social interests tend to argue, however, that WTO members, who may all be approaching an issue from a narrow commercial or national self-interest point of view, may collectively fail to appreciate the global externalities of environmental problems. Hence, they argue that they should have full access to, and participation in, the WTO to ensure that the externalities are addressed.

Given these significant weaknesses, the WTO appears to have reached its level of competence and will be unable to support the international integration of food-safety and environmental-protection regulations and thus deal with social regulatory barriers as issues become increasingly socio-economic in nature and calls for protection emerge from non-traditional sources.

The question then arises, should the agreements be broadened in order to address these concerns? Or should new agreements under the WTO be established? An innovative research paper argued that neither the SPS nor the TBT Agreements should be renegotiated to incorporate these concerns (Perdikis *et al.*, 1999). Instead, it was proposed that a separate Agreement on Trade Related Aspects of Consumer Concerns (TRACC Agreement) be negotiated within the WTO framework. Further, the international standards for this agreement would be established by a professional, social-science-based institution, the Commission on Consumer Issues and Trade.

The problem with the WTO framework is, however, much deeper than the limitations of the SPS and TBT Agreements. In fact, it is the traditional trade model supporting the scientific-rationality regulatory paradigm and the regulatory-competition approach to integration that limits the WTO as a framework capable of facilitating social regulatory integration, and it is unlikely that the WTO can shift enough within this paradigm to accommodate the broader concerns. To incorporate broader social dimensions would require deeper integration, which is not an objective of the traditional trade model.

Therefore, the WTO cannot effectively facilitate the international integration of GM-crop regulations. The key issue is that, when fundamental framework regulatory principles have not been established, the WTO cannot develop top-down regulatory rules for risk assessment and risk management. Instead, dominant jurisdictions must establish their own approach, subject to endogenous political-economy factors, and then seek to integrate that approach with those of other jurisdictions. In other words, while the WTO can remain effective at administering and monitoring regulatory rules, it cannot make the regulatory framework, as this needs to be done through bottom-up intergovernmental regulatory integration. Every effort should be made by all members to avoid bringing social regulatory-barrier disputes to the WTO. In this way, the WTO can continue its valuable trade-liberalization efforts on behalf of members, without having to risk its credibility and legitimacy by making risk-management judgements against the social-regulatory approaches of members.

5.3.2 GM-crop regulatory integration

Given the conclusion that there are no top-down WTO solutions for dealing with regulatory regionalism resulting from social regulatory barriers to GM crops, it is crucial to discuss how regulatory integration might then occur. In Chapter 1 it was proposed that, in respect of social regulatory barriers, a regulatory jurisdiction faces three choice parameters for regulatory integration: level of integration, depth of integration and strategy of integration (Fig. 5.3). These parameters may now be applied to the specific challenge of social regulatory barriers to GM crops, and they will be assessed in this order below.

A. Level of integration

This study proposes that the appropriate level of integration is the bilateral transatlantic level and not a top-down global or multilateral level. There are several reasons for this. First, North American and European agricultural biotechnology firms have the greatest capacity for developing and commercializing GM crops. As a result, both regions have had to develop regulatory oversight in this rapidly expanding sector. Secondly, with high levels of income, both regions have implemented stringent income-elastic social regulations for food safety and environmental protection. Thirdly, transatlantic trade tensions play a pathfinder role for solving frontier market-access issues, such as social regulatory barriers, which must be addressed at the transatlantic level before they can be effectively multilateralized (Woolcock, 1998). This is the case because, although the US and the EU markets can normally compel other countries to unilaterally converge their regulations in order to gain market access, when the differences are transatlantic these two regions cannot compel one another

1. Integration Level

I. Global/multilateral	II. Regional/bilateral or plurilateral

3. Integration strategy

2. Integration depth		Regulatory competition	Regulatory coordination
	Shallow 'economic' integration	A	B
	Deeper 'social' integration	C	D

Fig. 5.3. Integration parameters facing a regulatory jurisdiction.

to converge and thus must embark on a dedicated integration strategy. Fourthly, social regulatory barriers have already had a significant trade impact in respect of the transatlantic dispute over hormone-treated beef.

Agricultural trade issues have been, and will continue to be significant impediments to transatlantic trade negotiations (Kramer, 1989; Phillips, 1991; Wigan, 1998; Scher, 1999). It has been argued that agricultural competition has been at the heart of all EU–US agricultural trade disputes, not food safety or environmental protection (Coffey, 1993; Lister, 1996). Given the competition issues in both regions, agriculture will remain a highly protected sector, despite recent attempts to introduce multilateral trade discipline through the Agreement on Agriculture of the WTO.

The discussion on the Cartagena Biosafety Protocol in Chapter 4 indicated the difficulties associated with international regulatory integration when there is no transatlantic consensus. Indeed, in the absence of a consensual transatlantic regulatory approach, it is virtually impossible to develop an international regulatory approach. Instead, international regulations for GM crops are a fragmented collection of standards, codes, guidelines and recommendations. It appears that all other issues surrounding agricultural biotechnology are stalled until the transatlantic regulatory regionalism is addressed. Unfortunately, this includes the crucial North–South capacity gap, international concerns regarding the transboundary movement of living modified organisms and biopiracy. In this sense, addressing the transatlantic regulatory regionalism is time-sensitive and vital to the economic development of less developed countries.

In order to understand the prospects for transatlantic regulatory integration, it is useful to discuss the background to the current transatlantic relationship. Traditionally, transatlantic economic relations have been regarded as a part of the larger strategic agenda of managing the global economic order to promote security and stability (Frost, 1998). From this perspective, the 'low politics' of economic relations are inherently linked to the 'high politics' of global security and stability, as increased trade produces increased interdependence and prosperity, in turn producing economic and social stability and security. Indeed, Article 2 of the 1949 North Atlantic Treaty Organization (NATO) Charter commits members 'to seek to eliminate conflict in their economic policies' and encourages 'economic collaboration between any or all of them' as part of the broader strategic objective (Frost, 1998). In the past two decades, there have been several proposed initiatives between both Canada and the EU and the EU and the USA designed to enhance transatlantic economic relations and to establish mechanisms for dealing with market-access hindrances. Is this link between low and high politics still a factor? It has been argued that there has been a decoupling of the high and low objectives (McCormick, 1998; Scherlen, 1998). In fact, despite these efforts to reinforce the transatlantic relationship, recent transatlantic trade tensions involving bananas and hormone-treated beef have occurred during the United Nations (UN) action in the Gulf and the NATO action in Yugoslavia, suggesting that the link between the 'high politics' of global security and stability and the 'low politics' of transatlantic economic relations has become very weak.

CANADA–EU INTEGRATION. In 1976, Canada and the EC completed the Framework Agreement for Commercial and Economic Cooperation, designed to facilitate bilateral consultation on trade issues.[14] In fact, this agreement was the EC's first cooperation agreement with an industrialized nation. This was followed in 1990 with the Transatlantic Declaration on EU–Canada Relations. Similarly to the 1976 Framework Agreement, the declaration was focused on facilitating bilateral consultation in order to promote economic relations.

In Ottawa, Canada, on 17 December 1996, Canada and the EU adopted the Joint Political Declaration and Action Plan, which consists of four parts: economic and trade relations; foreign-policy and security issues; transnational issues; and fostering further links. This Action Plan was an attempt to reinforce bilateral economic and political relations after several conflicts, such as the beef-hormone dispute and the Atlantic fisheries dispute between Canada and Spain. With the four-part agenda of the Joint Action Plan, it appeared that Canada–EU economic relations were recoupled with larger strategic issues.

Yet, as the Joint Action Plan was implemented, it was the trade and economic-relations component that received the most attention. For instance, the 1998 EU–Canada Trade Initiative (ECTI) established an agenda for bilateral economic and trade relations through agreements on mutual recognition, equivalence and regulatory cooperation in areas of technical barriers to trade. The recent EU–Canada Veterinary Agreement 1999 is an example of a bilateral agreement based on the mutual recognition of sanitary measures applied in both Canada and the EU. The ECTI is held to be the vehicle for addressing bilateral trade friction associated with agricultural biotechnology products. Therefore, despite the economic and political focus of the 1996 Action Plan, it is the economic component that has received the most attention, suggesting a decoupling of low politics from high politics.

EU–US INTEGRATION. With respect to EU–US bilateral relations, the transatlantic tensions associated with agricultural liberalization

[14]EU–Canada: A Solid Economic Relationship June 1999 (www.europa.eu.int/comm/dg01/eucanen4.htm) and EU–Canada: Historic Friends June 1999 (www.europa.eu.int/comm/dg01/eucanen1htm)

during the Uruguay Round of multilateral trade negotiations gave rise to three EU–US proposals. The first was the broad Transatlantic Declaration on EU–US Relations in 1990, which identified shared, long-term goals and linked the economic relations to political and security relations. The second initiative was the European Economic Community (EEC)–US Hague Summit in November 1991, followed closely by the third initiative, the EEC–US Declaration in West Berlin in December 1991. All three initiatives were aimed at breaking the agricultural impasse at the multilateral trade negotiations and together were successful efforts in economic diplomacy (Coffey, 1993).

In 1995, there were three EU–US proposals to enhance transatlantic relations. The Transatlantic Treaty proposal involved the establishment of a security, political and economic arrangement, effectively strengthening the link between economic relations and political and security relations. This represented deep integration with formal and binding rules.

There was also a proposal for a Transatlantic Free Trade Agreement (TAFTA). Supporters argued that there were three significant benefits from a TAFTA. First, the EU and the USA have compatible markets with similar demands, tastes and social and political values (Gaster and Prestowitz, 1994). Secondly, a TAFTA would represent an inter-regional trade-liberalization strategy, removing the limitations of regulatory regionalism and enhancing multilateral trade (Preeg, 1996). Thirdly, efforts to enhance trade liberalization between the two globally dominant regions would compel other protectionist states or regions to liberalize (Yeutter and Maruyapma, 1995). Critics argued that a TAFTA would create transatlantic regionalism, built on similar income levels and social values, which would both antagonize other countries and regions and harm the prospects for multilateral trade liberalization (Bergsten, 1996).

Thirdly, in 1995, the New Transatlantic Agenda (NTA) was proposed, focused on promoting global peace, stability, democracy and economic development, while both contributing to the expansion of world trade and enhancing transatlantic economic relations. The NTA was a policy document with no binding discipline and lacking the power to compel both sides to address, in a meaningful way, the most contentious issues facing transatlantic economic relations (Wigan, 1998).

One particular initiative of the NTA was to establish the New Transatlantic Marketplace (NTM). Specifically, within the NTM initiative was a proposal to establish a US–EU biotechnology task force to encourage dialogue between the US Food and Drug Administration (FDA) and the EU Directorate General (DG) III Industry. It is important to note that, as will be discussed in more detail below, these two regulatory agencies agree on many fundamental principles about biotechnology regulations. The regulatory regionalism in respect of GM crops is the result of the dominance of DG XI Environment in EU regulations, because DG XI has a very different regulatory philosophy from that held by the FDA and DG III.

The three EU–US transatlantic proposals in 1995 began with attempts to develop formal and binding integration across economic, social and political issues and ended with attempts simply to outline a non-binding framework for shallow integration across primarily commercial or economic issues – a shift from high to low politics.

Coinciding with the NTA were three separate initiatives: the Transatlantic Business Dialogue (TABD), the Transatlantic Consumers Dialogue (TACD) and the Transatlantic Environmental Dialogue (TAED). These initiatives sought to bring together relevant stakeholders in the three broad policy areas, including government, industry and non-governmental organizations, in order to address transatlantic issues. For instance, the TABD, established in Seville, Spain, in November 1995, is an informal group of companies and industry associations who develop joint EU–US trade-policy recommendations for dealing with regulatory barriers to trade from an industry perspective. The TABD's 1997 Communiqué (Rome) noted that 'our regulatory agencies can no longer continue to function solely on the basis of national considerations' and it urged the respective administrations to pursue regulatory harmonization (Zampetti, 2000). However, it has been argued that TABD's attempts at resolving regulatory structural market-access issues have had only limited success (Frost, 1998).

The most recent EU–US transatlantic initiative has been the Transatlantic Economic Partnership (TEP), adopted at the EU–US London Summit of May 1998. The TEP is a bilateral agreement to discuss key transatlantic issues, ranging from market-access barriers in goods and services to environment and competition law. As the name suggests, the TEP focused largely on the integration of commercial or economic issues. Similarly to the NTA, the TEP is only a framework agreement; it is not a binding commitment for the EU or the USA.

In the USA, the relevant agencies are the US Trade Representative (USTR), US Department of Agriculture (USDA), FDA and the Environmental Protection Agency (EPA). In the EU they are DGs I (Foreign Affairs: Trade), III (Industry), VI (Agriculture), XI (Environment) and XXIV (Consumer). Under the TEP, the EU and USA are committed to biannual political summits, supported by more frequent cabinet-level meetings, designed to direct the various sectoral consultations. Then the TEP Steering Group monitors the daily efforts of the TEP sectoral consultations.

It has been argued that this administrative approach has been proposed because dealing with transatlantic issues requires a coordinated approach across many sectors, creating the political momentum to move forward, despite difficulties that may emerge in respect of any one regulatory issue (Zampetti, 2000).

An important aspect of the TEP was that it was meant to provide global leadership in agenda setting for multilateral trade negotiations. Efforts involve identifying shared priorities and objectives, establishing regular dialogue and improving cooperation among scientists and regulators in order to enhance regulatory cooperation. At the same time, the TEP also attempts to increase participation in trade policy for domestic interest groups, such as consumer and environmental organizations, who have long felt left out of the discussions on transatlantic integration. In this sense, the integration strategy of the TEP is one of regulatory coordination.

Agricultural issues associated with biotechnology are an identified priority area for bilateral consultation (Transatlantic Economic Partnership, 1998). Both the EU and the USA support the need for studying the regulatory structural market-access barriers to agricultural biotechnology products (Wigan, 1998; Schwarz, 1999). In order to address the trade friction associated with regulatory regionalism, the objective was to establish a transatlantic sectoral consultation group with a twofold mandate:

- to monitor progress of the dialogue on the various technical issues carried out in existing groups, and to take into account their potential trade effects with the objective of reducing unnecessary barriers to trade;
- to seek to increase and enhance scientific and regulatory cooperation and information exchange and promote transparency and information to consumers.

(Transatlantic Economic Partnership Action Plan: 14)

Further, it was recommended that the regulatory divergence be studied through a pilot project monitoring simultaneous applications in the EU and in the USA. While the target date for this project was the end of 1998, the sectoral consultation group met in February 1999 and was unable to launch the project, perhaps because at that time transatlantic trade disputes associated with bananas and hormone-treated beef were on the agenda.

To summarize, both Canada–EU and EU–US efforts at enhancing integration have eventually adopted an approach of technical, low-politics consultations among technocrats. There are two problems with this approach. First, according to Frost (1998), regulatory integration needs a high-level political push. Yet transatlantic economic relations in the past decade have steadily shifted from 'high politics' to 'low politics'. As high politics, national economic interests were subordinate to the larger strategic aims. Now, transatlantic economic relations are decoupled from the strategic aims and do not have the same political encouragement for integration; consequently, transatlantic trade frictions have grown. A trade war tends to elevate transatlantic economic relations to high politics, but generally not in a beneficial way. Disputing countries become adversaries, not partners, and the final outcome is forced convergence, which creates winners and losers rather than mutual beneficiaries. It has been noted that, given the market-access difficulties faced by Monsanto in gaining approval for a GM variety of maize, the US Clinton Administration pressured the British government to approve this variety (*Independent*, 1999q). Such an application of 'high political' pressure is not constructive either for transatlantic economic relations or for the objective assessment of the safety of GM crops, and it should always be avoided. Secondly, populist issues are not congruent with the traditional trade-diplomacy approach of closed-door consultations among trade-policy experts. Without transparency and broader participation, a sceptical public will dismiss any technical regulatory integration decisions as insufficient. Instead, transatlantic regulatory integration must balance national-competitiveness demands for technological progress with the populist concerns and fears driving calls for technological precaution.

B. Depth of integration

Given the dominance of the transatlantic level in dealing with social regulatory barriers to GM crops, the next integration parameter is identifying the appropriate depth of regulatory integration. This is, essentially, a choice between shallow economic integration and deeper social integration.

The discussion of the development of GM-crop regulations has revealed that there are significant social issues beyond just economic issues, which have crucially influenced the regulatory development process. As a result, integration must be deep enough to address the social issues and concerns driving the regulatory approaches. In other words, social rationality matters and, in dealing with social regulatory barriers to GM crops, trade diplomacy cannot sustain the traditional divide between economic and social integration. Instead, it must endeavour to include crucial social interests fully in the trade negotiations. Failure to go 'deeper' immediately antagonizes social interests and undermines the domestic support for transatlantic integration based only on the international economic-integration approach of traditional trade diplomacy.

C. Strategy of integration

Once the integration level and depth are determined, it is crucial to identify the relevant strategy. Recall that the choices are essentially between regulatory competition and regulatory coordination. The discussion on the traditional trade approach has revealed significant social concerns with regulatory competition.

There are two types of regulatory coordination strategies to consider: harmonization and mutual recognition. Harmonization represents deep integration leading to the establishment of identical regulations in different sovereign states. Under this coordination strategy, national regulatory differences are systematically adjusted to be congruent with regulatory approaches in other states (Wiegele, 1991). Harmonization has several benefits. First, from an economic perspective, it can eliminate market fragmentation based on divergent regulatory approaches and, in essence, level the playing-field, reducing trade tensions and enhancing economies of scale. Secondly, from a social perspective, harmonization is congruent with addressing both food-safety and environmental-protection regulatory externalities.

There are, however, crucial limits to the harmonization of national regulations, because of important subsidiarity forces. Harmonization requires national regulatory compromise. From an economic perspective, the potential for agricultural biotechnology in determining future national comparative advantage pressures authorities to retain full control over regulations. Proposed regulatory compromises might not be in the interest of national competitiveness. Indeed, this is exacerbated by the current North American commercialization lead in agricultural biotechnology over Europe. From a social perspective, regulatory compromise might conflict with domestic consumer or environmental-protection concerns. Indeed, it would be unwise for national governments to concede regulatory authority or compromise in politically sensitive areas, such as consumer and environmental protection.

Additionally, from an institutional perspective, since agricultural biotechnology is a horizontal issue, many domestic government departments and agencies have a regulatory role to play. There are two complications associated with this. First, securing a package of harmonized regulations within these national regulatory jurisdictions would be difficult, let alone the complications of then harmonizing the regulations with another state. Secondly, in an era of government downsizing and streamlining, departments and agencies may be unwilling to concede regulatory authority over portfolios that carry political weight. Therefore, transatlantic regulatory coordination must recognize that crucial subsidiarity forces drastically limit the efficacy of harmonization as an appropriate coordination strategy.

The second type of regulatory coordination strategy is mutual recognition. This strategy allows for divergence in regulatory approaches and does not pursue the same depth of integration. For instance, states can recognize the hypothetical risk assessment of other states, but then conduct their own risk-management procedure. Regulatory coordination is achieved when states agree that, although their regulatory approaches differ, the approaches achieve equivalent outcomes. Mutual recognition limits the need for regulatory compromise.

Yet mutual recognition has some limitations as well. First, it is limited when regulatory divergence is significant, because equivalency requires at least some degree of similarity in approach. It has been argued that, where there are serious disagreements, such as those dealing with risk analysis, even identifying basic shared objectives is difficult, jeopardizing integration (Gatsios and Holmes, 1998). It has been revealed that fundamental principles for regulating GM crops have not been established, making equivalency at even a minimum level extremely difficult. Secondly, mutual recognition may fail to address regulatory externalities, such as consumer and environmental protection. Two states may mutually recognize their respective approaches and yet neither approach may address the externalities.

Given the benefits and limitations of both harmonization and mutual recognition, it may be asked: what type of coordination can result in transatlantic regulatory integration that addresses regulatory externalities while simultaneously remaining compliant with powerful subsidiarity forces?

The literature on European integration has dealt extensively with regulatory coordination among national governments unwilling to harmonize regulations or incapable of doing so. To overcome this problem, a European 'new approach' to regulatory coordination has been adopted (Woolcock, 1996). The new approach is the result of a 1985 White Paper on the Internal Market, where it was proposed that 'harmonization could be accelerated if the European Council withdrew from the details of regulation' (European Commission, 1985). Instead, the Council would agree to a framework and the relevant service of the European Commission would develop the regulatory approach. It was decided that, beyond essential minimum requirements, 'politics' produces bad regulations, because it complicates regulatory development.

Under the new approach, essential minimum requirements (EMRs) are established for a particular regulatory issue. These are set at an acceptable level of oversight, creating a regulatory floor and preventing a regulatory race to the bottom. The EMRs are set either by the Council of Ministers or by the ECJ. At the level of EMRs, some degree of harmonization is necessary (Majone, 1993). Establishing EMRs is essentially a normative exercise of identifying the acceptable social objectives that become the regulatory parameters. On top of the EMRs is a coordination strategy of mutual recognition.

Is the new approach to European regulatory coordination an appropriate approach for the transatlantic regulatory coordination of agricultural biotechnology? Perhaps, but there remain significant challenges. Unlike the European example, where countries broadly agreed upon the minimum public interest, an important problem with agricultural biotechnology is the harmonization of essential minimum requirements. As will be discussed in Chapters 6 and 7, there are significant differences in the respective transatlantic positions on the framework debates, let alone the specific debates. Also, in the case of transatlantic trade, there is no court of justice or council of ministers to impose EMRs if none can be established.

In order to identify the appropriate transatlantic regulatory coordination strategies for GM crops, it is necessary to understand the domestic political-economy factors of the North American and the EU regulatory approaches. In Chapters 6 and 7, respectively, the regulatory approaches will be characterized according to the framework and specific regulatory debates discussed above.

5.4 Conclusions

In Part II, Chapters 3, 4 and 5, two conceptual models for analysing social regulatory barriers to GM crops have been developed: the model characterizing a regulatory approach within a jurisdiction (Table 5.10) and the model characterizing the regulatory integration strategy employed (Figure 5.3). The crucial conclusion is that the regulatory approach, which is a function of endogenous political-economy factors, determines the prospects for regulatory integration. With respect to GM crops, the scientific- and social-rationality paradigms support divergent regulatory trajectories, characterized by seven framework debates and five specific regulatory debates. This creates regulatory instability, reducing the certainty and predictability of regulatory approaches within jurisdictions and hindering the potential for regulatory integration between jurisdictions. Economic interests support the international economic-integration approach of scientific rationality exemplified by the international trade regime. Social interests support the international social-integration approach. Due to this conflict, there is no obvious international institution, including the WTO, to achieve regulatory stability and integration. Instead, the most important level of regulatory integration is the transatlantic level, where the integration must be capable of meeting the social aspects of domestic regulations.

Part III
Transatlantic Regulatory Regionalism

The objective of Part III is to examine the transatlantic regulatory regionalism associated with the development and commercialization of genetically modified (GM) crops, in order to identify the prospects and limitations for regulatory integration.

Both the North American and the European regulatory approaches are based on the risk-analysis framework. In fact, until 1990, they were essentially on the same regulatory trajectory, with largely congruent positions on the various regulatory principles, reflecting a scientific-rationality paradigm. By 1990, however, endogenous political-economy factors resulted in a shift of European Union (EU) regulation away from this common trajectory and towards a significantly divergent framework, more closely reflecting the social-rationality paradigm.

The result has been the creation of transatlantic regulatory regionalism and regulatory market-access barriers facing GM crops. Essentially, GM varieties of agricultural crops approved in the North American market-place have not gained similar approval in the EU. Instead, applications have faced significant regulatory hold-up and rejection. For example, a GM variety of flax-seed, CDC Triffid, approved for commercial release in Canada in 1997, was neither approved nor rejected in Europe. Instead, the application remained in a regulatory limbo (McHughen, 2000). Trade concerns arise because about 90% of Canadian flax is exported and the EU is the largest export destination. Although approved in Canada, Canadian producers and distributors voluntarily refused to plant the GM variety commercially, because, without an EU-accepted identity-preservation production (IPP) system, exports to the EU could not be guaranteed as free of the GM variety approved in Canada but not in the EU. All

Canadian flax exports have been banned. In fact, this is exactly what has happened to Canadian canola/rape-seed exports and US maize exports to the EU in the period 1997–1999. In Canada, a total of seven GM varieties of canola were approved for commercial release and are widely planted. Yet only four of the seven had been approved in the EU, while the remaining three applications were held up in the EU's regulatory approval system. Again, without an IPP system, segregation of the approved from the non-approved varieties could not be ensured, nor could segregation of GM from non-GM varieties. As a result, Canadian exports of canola to the EU fell dramatically (Phillips and Isaac, 2001). In the USA, it has been estimated by the US Department of Agriculture that the transatlantic regulatory regionalism has resulted in over US$200 million in lost maize sales.

In order to understand the divergent regulatory approaches causing the transatlantic regulatory regionalism, each jurisdiction will be characterized according to the scientific- and social-rationality paradigms presented in Chapter 5. This includes identifying the competitive position, the traditional regulatory role of the state, the interests involved in regulatory development and the structure and stability of the regulatory regime, along with an assessment of the regulatory integration strategy that follows from the regulatory approach. The North American regulatory approach will be characterized in Chapter 6, while the European approach will be characterized in Chapter 7.

6 North American Regulatory Approach

6.1 Introduction

The objective of this chapter is to characterize the North American regulatory approach to agricultural biotechnology products by examining its development in the USA and in Canada. This includes identifying the public initiatives to promote biotechnology, as well as the efforts to regulate agricultural biotechnology. In this chapter is also an examination of the degree of Canada–US regulatory integration made possible because of the similarities in regulatory approach.

Prior to discussing the North American approach to regulating genetically modified (GM) crops, it is useful to understand the social context of agricultural production. Unlike in many parts of Europe, agricultural land in North America is not also 'wilderness'. Instead, the agricultural sector is primarily an industry for making foodstuffs and does not double as the protector of the countryside. This is not to say, however, that North American agricultural producers are not themselves concerned with the environmental impact of their activities and, indeed, many make great efforts to adopt sustainable practices. It is only to suggest that the broader public does not envisage a special 'multifunctionality' role for the farming community to the extent found in Western Europe (to be discussed in Chapter 7).

6.2 Agricultural Biotechnology in the USA

The USA is the global leader in biotechnology capacity in general and is the dominant country in agricultural biotechnology in particular (Office of Technology Assessment, 1991; James, 1999). In essence, the

US experience with biotechnology has significantly influenced the promotion and regulation of biotechnology in the rest of the world.

6.2.1 US Promotion of agricultural biotechnology

The global dominance of the USA in biotechnology capacity has occurred without strategic industrial-targeting programmes. It is argued that the USA has lacked, and indeed continues to lack, a comprehensive industrial development strategy aimed at building biotechnology capacity (Wiegele, 1991). Instead, public policies are more broadly targeted at the development of all industrial initiatives employing knowledge-based, innovative, advanced technologies. Public funding through the federal government is available to assist basic, non-commercial research in these areas in collaboration with universities, research institutes and private companies. Funding for commercially orientated research is generally left to the financial market. In fact, it has been argued that the US regulatory framework has had more to do with the private-sector-led commercialization of agricultural biotechnology products than any public policies of industrial targeting (Lavoie and Sheldon, 1999). For example, in 1991, the President's Council on Competitiveness recommended two key principles of US biotechnology regulations that assisted technological progress. The first was a focus on product standards, rather than technology or process standards, while the second was to regulate under existing vertical or sectoral regulatory departments. Both of these principles are key features of the scientific-rationality approach encouraging technological progress. Further, on 30 November 1999, the US-based Alliance for Better Foods held a news conference in Seattle, Washington, to publicly support agricultural biotechnology and the US regulatory system. Most notable was a statement of support for agricultural biotechnology, GM crops and the US regulatory approach undersigned by 330 scientists (*AgraFood Biotech*, 1999b). The importance of these two examples is that the regulatory approach has been explicitly linked to the promotion of biotechnology through establishing stable and predictable regulatory rules that encourage technological progress.

6.2.2 The US regulatory system

Due primarily to the global-capacity lead, the USA has led the rest of the world in developing a regulatory base for biotechnology products. The initial regulatory approach was driven by scientists many years before there were any market-ready commercializations of biotechnology-based products. The initial approach was based upon the desire of these scientists to proceed with caution in genetic-modification techniques.

With respect to the traditional US regulatory role, it has been argued that regulatory intervention in the USA follows a regulatory-independence approach (Majone, 1990). Congruent with the economic perspective, the regulatory-independence approach prescribes government regulatory intervention only in reaction to market failure,[1] where regulations seek to ensure market efficiency or effectiveness. For example, a lack of sustained technological progress would be interpreted as failure to adequately reward innovation, resulting, perhaps, in the development of regulatory policies to protect intellectual property. Accordingly, the market decides the effectiveness of regulations based on their ability to facilitate technological progress while delivering an acceptable level of safety. Regulators then have a mechanistic task and must be independent of possible capturing interests. With respect to modern biotechnology, it is quite appropriate to suggest that US regulations have been pro-competitive and focused on removing market failure to enhance market efficiency and effectiveness.

This traditional regulatory role has been congruent with technological progress and the development of a scientific-rationality approach to US biotechnology regulations. In June 1973, the Gordon Conference on Nucleic Acids participants recommended to the US National Academy of Sciences and the US Institutes of Medicine that greater attention be given to possible biohazards from new viral strains emerging from laboratory research (Wiegele, 1991). In response, the National Academy of Sciences and the Assembly of Life Sciences of the US National Research Council appointed a panel of expert scientists to examine the biohazard concern raised by the Gordon Conference participants. The ensuing report recognized that there is a potential risk (either a hypothetical or speculative risk), rather than a recognizable risk, from biotechnology, but nevertheless encouraged a voluntary moratorium on some types of experiments until the international scientific community could discuss the risks (Berg *et al.*, 1974). This recommendation remains consistent with the scientific interpretation of the precautionary principle, where 'precaution' was linked to the gathering of further scientific-risk information for some experimental applications, while for other applications there was simply too much uncertainty to perform even a hypothetical risk assessment, defaulting instead to a moratorium.

[1]Market failure: 'The inability of a system of private markets to provide certain goods either at all or at the most desirable or "optimal" level ... The existence of market failure is usually held to provide a case for collective or government action to improve allocative efficiency' (Pearce, 1996).

Following the report's recommendations, in February 1975 the National Academy of Sciences convened the International Conference on Recombinant DNA, known as the Asilomar Conference, in Pacific Grove, California. The Asilomar Conference focused on two points. The first was how to minimize 'potential' risk and the second how to involve the public appropriately. With respect to minimizing risk, the participants agreed on the establishment of a two-stage graduated scale for assessing risks: low risk and high risk. It has been argued that the low risks were hypothetical risks, while the high risks were speculative risks (van den Daele *et al.*, 1997). With respect to involving the public, the participants recognized both the presence of a significant information gap and the adverse effects undue public fears could have on important and vital research. Therefore, conference participants were keen to establish guidelines that the public would consider appropriate and acceptable. Further, while the guidelines were established to deal with hypothetical risks, there was awareness that speculative risks – regardless of their basis – could generate concerns and fears that would, in turn, jeopardize research efforts. In this sense, the scientific community was aware that speculative risks could not be ignored.

Also in 1995, the Recombinant DNA Advisory Committee (RAC) was established by the National Institutes of Health (NIH). The RAC played an influential role in setting guidelines and recommendations for research using genetic-modification techniques (Ager, 1990). In 1976 the RAC compiled the first guidelines related to recombinant DNA (rDNA) research, which were released by the NIH as the *NIH Guidelines for Research Involving Recombinant DNA Molecules.* These guidelines were focused only on controlled research, not on experimental or commercial environmental release. There were two important aspects of the guidelines. First, it was argued that there was no significant risk inherent in the use of biotechnology requiring technology-based regulations. Secondly, it was recommended that the degree of regulatory oversight should vary, based both on the degree of research containment and on the scientifically determined hypothetical risk. The guidelines reflected a commitment to both the principle of 'substantial equivalence' and the scientific interpretation of the precautionary principle – the base for the US regulatory structure.

Within 5 years of the Asilomar Conference and the NIH guidelines, most scientists had reassessed their initial precautious judgements about the safety hazards of rDNA research, because of continued experimentation and accumulating risk information. As a result, there was substantial support for a relaxation of the guidelines in light of this information (Ager, 1990; Wiegele, 1991). This indicates a discernible shift from hypothetical risks to recognizable risks, made possible by the scientific interpretation of the precautionary principle.

Although imposing initially restrictive, precautionary guidelines, it allowed for the research and development (R&D) necessary to better understand the hypothetical risks, which in turn led to a relaxation of the guidelines, based on known or recognized risks, in turn fostering further R&D and the accumulation of risk information.

Arguably the most important regulatory development in the USA was the publication by the National Academy of Sciences of the National Research Council's 'Redbook' on regulating the risks of advanced technologies (National Academy of Sciences, 1983). The important aspect of this publication was that it set out a new regulatory paradigm based on the risk-analysis framework. This framework significantly shaped the US regulatory approach to modern biotechnology (Mackenzie, 1993). The Redbook identified the now familiar risk-analysis trilogy of risk assessment, risk management and risk communication (see Chapter 5).

With the growth of R&D in agricultural biotechnology, pressure began to mount for guidelines to deal with the imminent environmental release of genetically modified organisms (GMOs). There were two guidelines that responded to this pressure. In 1985, a coordinating committee, the Biotechnology Science Coordination Committee (BSCC) was established by the President's Office of Science and Technology Policy (OSTP) and staffed by senior academic and industry scientists (Wiegele, 1991). Through the OSTP, the BSCC published the first guideline to deal with environmental release, the *Coordinated Framework for Regulation of Biotechnology* (Office of Science and Technology Policy, 1986). The second guideline was the National Research Council's *Field Testing Genetically Modified Organisms: Framework for Decisions*. These guidelines were very similar and they both concluded that, although some biotechnology applications would create novel products with novel hazards, other applications would not, or, in other words, some GM crops are substantially equivalent to conventional varieties. Both guidelines supported the application basis of the scientific-rationality paradigm by suggesting that there were no significant risks inherent in the application of biotechnology *per se*; hence, they did not support new technology- or process-based regulations. It has been argued, however, that these conclusions should not be interpreted as suggesting that GM crops are not different from non-GM crops; indeed, they are different, but not in the risk or hazard that they pose (MacKenzie, 1993). Instead, both guidelines suggested that additional regulations for GM crops combined with interagency coordination were necessary, but as part of the existing vertical regulatory oversight.

In 1990, the BSCC was replaced by the Biotechnology Research Subcommittee, an indication of the 'permanence' of the biotechnology issue on the US industrial landscape. Further, as previously discussed, in 1991, the President's Council on Competitiveness recommended two key principles of US biotechnology regulations: first, to focus on product standards and, second, to regulate under vertical or sectoral existing regulations. These principles were viewed as necessary to ensure the safety of biotechnology-based products, while simultaneously enhancing the competitiveness and commercialization lead of the US biotechnology industries.

Clearly, since 1973, US biotechnology regulations have been on a scientific-rationality trajectory, employing safety- or hazard-based definitions of risk, the substantial-equivalence principle and the traditional burden of proof. The result has been a product- or application-based regulatory approach under the jurisdiction of existing, vertical departments and agencies and employing the scientific interpretation of the precautionary principle.

Three agencies currently share responsibility for the regulation of GM crops. The first, the Animal and Plant Health Inspection Service (APHIS) of the US Department of Agriculture (USDA), is responsible for environmental assessments of plant risk, for issuing Environmental Use Permits (EUPs) for field-testing and for regulating the importation and interstate movement of GM plants. In this sense, APHIS does have a process focus explicitly on those crops derived from GM techniques. APHIS examines all organisms and products altered or produced through genetic engineering that are or have the potential to be plant pests. The APHIS regulatory procedures have been simplified twice. In 1993, six GM varieties were 'delisted', meaning that they only required notification for field-testing, not an approval or an EUP. This action was based on the confidence of accumulated field-trial experience in the USA. Under the second simplification, adopted to speed the application review, APHIS regulations provide for a petition process where the proponent can request that their product be granted a non-regulated status. The proponent submits the necessary evidence to prove that the GM plant does not pose a plant-pest risk. APHIS reviews the evidence and, if approved, the plant product (and all its offspring) no longer requires APHIS review for interstate movement or environmental release in the USA. The necessary evidence involves a scientific risk assessment.

The US Environmental Protection Agency (EPA) is responsible for the environmental release of both bioengineered pesticides and bioengineered plants with pesticidal characteristics, including insecticidal varieties, such as *Bacillus thuringiensis* (Bt) maize and Bt cotton varieties. In this, similarly to APHIS, the EPA also has a process-based focus on GM crops with pesticidal characteristics, which are subject to EPA assessments, considering the impacts upon the target organisms and non-target organisms, including humans. Similarly, for GM varieties of herbicide-resistant crops, the EPA establishes tolerances for herbicide residues. Although a process-based focus is initially employed to determine crops within the EPA jurisdiction, a scientific-rationality approach is used in the risk-assessment stage, employing principles such as hypothetical risks and substantial equivalence in order to determine novelty. Also, it has been argued that the Toxic Substances Control Act (TSCA), administered by the EPA, explicitly employs the scientific interpretation of the precautionary principle, since 'the language of the statute express Congress's clear intent that research and development activities involving small amounts of substances be exempt from regulation' (Rogers, 1990).

In 1999, the EPA's Ecological Effects Team announced that 'companies developing GM crops will face changes in testing requirements over the coming years' (*AgraFood Biotech*, 1999b). This reflects two trends. First, the increase in novel applications in GM crops necessitates new testing requirements, including more comprehensive scrutiny. Secondly, improvements in a predictive ecology framework for risk-assessment methods will better guide regulators to improve the accuracy and acceptability of environmental risk assessments. In other words, these changes must not be interpreted as a dramatic reversal in EPA policy. Instead, they are simply the natural evolution of regulations as GM crops move from production-trait varieties to both output-improved varieties and bioengineered commodities and as regulators gain more data on recognized risks and more experience with predictive ecology-based risk assessments.

The third institutional arrangement in the US regulatory system is the Food and Drug Administration (FDA) of the Department of Health and Human Services. The FDA has traditional responsibility over ensuring the safety of food and feed-use crops. It is important to note that FDA consultation is not mandatory but recommended prior to the market release of GM foods and feeds. Unlike the initial process focus of both APHIS and the EPA to identify crops that fall within their jurisdiction, the FDA employs a strict novelty focus in determining food products that require regulatory oversight or not. According to a 1992 FDA Statement, the focus of FDA oversight is on the novelty of food plants, not on the use of biotechnology *per se* (Food and Drug Administration, 1992), such that biotechnology-based products only come under direct FDA jurisdiction if they are determined to be a food additive according to the authority of the Federal Foods, Drugs, and Cosmetics Act (FFDCA). The 1992 Statement concluded, however, that transferred genes are not additives, because there is no *a priori* reason to conclude that transferred genetic material would create new hazards not present in conventional products (Koster and Balf, 1998). Accordingly, the FDA regulates the use of biotechnology under the substantial-equivalence principle – only if it is from a novel plant that is significantly different in structure, function or composition from plants currently used in food products. Many of the GM crops currently being developed for food use do not contain substances that are significantly different from those already in the diet and thus do not require the FDA's pre-market approval.

While FDA consultation is voluntary, there have been no GM products approved for commercial use that have not gone through FDA consultation. Simply, industry recognizes the important confidence boost that an FDA endorsement gives to products. Also, FDA consultation is *de facto* standard, because of US product-liability laws. From the firm's perspective, food-safety risk is minimized and, consequently, commercial risk is minimized through FDA consultation, as such a consultation limits liability risk. Many GM supporters, however, recognize that the FDA consultation is vulnerable to attack by critics, because of the fact that it is voluntary, despite the reality that the consultation is *de facto* standard. This has led some to argue that, as the consultation occurs anyway, it would be useful, from a public-concern perspective, to formalize the pre-market consultation in order to stem the potential criticisms.[2]

There is a horizontal regulation in the USA with oversight for environmental issues; hence, it is relevant for GM crops. The National Environmental Protection Act (NEPA) requires that all departments and agencies must perform an adequate environmental assessment on all activities falling under their jurisdiction that will have an impact upon the environment and whose impact is not already explicitly assessed according to a vertical regulatory approach. With respect to agricultural biotechnology, all deliberate releases for the field-testing or commercial planting of GM varieties is regulated under either APHIS or EPA oversight, limiting the application of the NEPA. The only time the NEPA could be used to additionally regulate GM crops is if it can be shown that the initial, vertical assessment for environmental impact was not adequate. The US federal regulatory approach is presented in Table 6.1 below.

There also exists state-level regulatory oversight in the USA (Rogers, 1990). Before commercialization, GM plants must also conform with standards set by state and federal marketing statutes, such as state seed-certification laws. Several states have enacted legislation regulating field trials, either requiring notification of the release of GM varieties (Hawaii, Illinois, Wisconsin) or requiring formal permits for trials (Minnesota and North Carolina). Nevertheless, once the product has been approved for unconfined release at the federal level, the individual states lose regulatory oversight and cannot prohibit the unconfined release of GM crops.

[2]Barach (1999) (see *AgraFood Biotech*, 1999a).

The regulatory decision-making process according to the US approach may be characterized as having narrow participation in the process. Experts are left to perform the hypothetical risk assessments, where the experts are either federal regulators or the risk-assessment personnel of the proponent GM-crop developer. Under the risk-management procedures, the decision-making is made within the relevant vertical regulatory jurisdictions. It has been argued that the decision-making power rests with 'a small cadre of scientists, venture capitalists and government bureaucrats, most of whom have a vested interest in the growth of the genetic engineering industry' (Rogers, 1990). Yet others have argued, on the contrary, that, despite the limited participation in the decision-making procedures, there is considerable opportunity for public comment on the draft policies at both the risk-assessment and the risk-management stages (Fuchs, 1993). In fact, in 1992, each of the three regulatory agencies, APHIS, EPA and FDA, published draft regulatory guidelines for public comment (APHIS, 1992; EPA, 1992; FDA, 1992).

An added dimension to US regulatory development is the regulatory evaluation by the Office of Management and Budget (OMB). The OMB assesses proposed US regulations and legislation to ensure that new regulations will generate the desired outcome, while limiting the political influence behind them. The OMB evaluations support a

Table 6.1. US agricultural biotechnology regulations.

Agency	Regulatory oversight	
Vertical legislation		
US Department of Agriculture (APHIS)	Plant pests, plants, veterinary biologics	Initially process-based to determine oversight, then scientific-rationality approach to the risk-analysis framework is employed
Environmental Protection Agency	Microbial/plant pesticides, new uses of existing pesticides, novel microorganisms	Initially process-based to determine oversight, then scientific-rationality approach to the risk-analysis framework is employed
Food and Drug Administration	Food (except meat, poultry and egg products), feed, food additives, veterinary drugs	Novelty-based
Horizontal legislation		
National Environmental Protection Act (NEPA)	Requires an assessment of environmental impact by the relevant regulatory agency on all activities that are not already assessed under vertical regulation	

scientific-rationality approach to regulating advanced technologies. In fact, the OMB approach does not support engaging all stakeholders in the risk-management process, because such involvement can potentially distort regulatory development beyond the goal of correcting market failure (Belzer, 1998). OMB's regulatory-evaluation framework is composed of four objectives: first, to identify the market failure used to justify the intervention; secondly, to analyse the impact of the identified market failure to determine both its presence and its significance in distorting market outcomes (e.g. impact on competition, equity, employment, public health, etc.); thirdly, to analyse existing federal regulations and legislation to determine whether amendments to existing frameworks are capable of addressing the identified market failure; and fourthly, to analyse all regulatory alternatives in order to determine the optimal regulatory approach. With respect to the risk-analysis framework, the OMB evaluation supports risk assessment based on hypothetical risks, focused on safety or hazards only, as well as risk management limited only to the goal of reducing or preventing risk, not the goal of assuaging public fears and concerns.

The US regulatory system was subject to an independent scientific review by the National Academy of Sciences in 2000 and 2001. A major focus of this review was not on human health, but on the 'environmental impacts associated with the commercialization of biotechnology derived plants and to provide guidance on how best to assess and mitigate risks' (*World Food Law Monthly*, 1999b). The focus on the environmental impacts is a direct effort to address the speculative risks raised by critics through a science-based analytical approach to environmental risk assessment. Moreover, this focus is consistent with the shift in EPA oversight, previously discussed, to a predictive ecology framework for assessing environmental impacts.

With respect to the mandatory labelling of GM crops and foods produced from GM crops, the US position flows from the regulatory positions identified above. For instance, the determination of substantial equivalence and the FDA's 1992 decision that GM material in itself is not an additive do not support the consumers' right to know justification for labelling. This is consistent with the traditional safety or hazard justification for mandatory labelling (Korwek, 1993). Further, it has been argued that US case law reveals that 'consumer interest alone is not enough' for mandatory labelling and that a production method that has no discernible impact on the safety or quality of a final product is not a sufficient justification either (Koster and Balf, 1998). The FDA policy on labelling irradiated meat products was, however, a mandatory labelling policy justified by the consumer's right to know. The key to this decision is found under the labelling section of the FFDCA, Section 403(a)1, which requires that mandatory labelling must reveal 'material facts' in order to prevent misbranding. Irradiation as a production method was considered to be a material fact (Koster and Balf, 1998). What is the difference between irradiation technology and GM technology? It appears that GM technology, which is used in the seed-development and not in the crop-production and the food-processing stages, escapes this context of material facts. If GM technology were applied while crops were growing or when they were being processed, then this would be a material production fact that consumers would have a right to know about. Instead, because of the substantial-equivalence principle, GM crops are slotted into the same production and processing system as same-use non-GM crops and there are no material differences in the processing procedures.

There is currently considerable pressure on the FDA's position on mandatory labelling. For instance, the FDA's policy on substantial equivalence is under legal challenge by the US-based International Center for Technology Assessment (ICTA), a non-governmental organization that monitors the social implications of technological innovations (*Economist*, 1999a). There are two important aspects of this challenge. First, the ICTA wanted the FDA policy on GM material altered so that GM material would be considered to be an additive and therefore subject to FDA pre-market approval. Secondly, the ICTA wanted a mandatory US labelling policy on foods produced from GM crops according to the consumer's right to know. The outcome of this challenge is currently unknown. Also, Democrat Representative (Ohio) Dennis Kucinich presented a bill to the US Congress calling for mandatory labelling of GM foods based on the consumer's right to know and requiring a 0.1% level of adventitious contamination.

As a result of these pressures, as well as the broader public concern, the FDA is consulting with the public on its labelling policy. This is an important development, because the FDA recognizes the need for regulations to move beyond just scientific parameters and to assuage public fears in order to prevent a collapse in consumer confidence. The outcome of these consultations is likely to be that the FDA will not adopt a mandatory labelling policy based on the consumer's right to know and instead it will develop legal standards for using a voluntary GM-free label.

With respect to the international integration of biotechnology regulations, the USA is involved in bilateral and multilateral negotiations both for the development of an international regulatory approach and for the enhancement of regulatory convergence between countries. Bilateral efforts include those taken under the North American Free Trade Agreement (NAFTA) and those under the Transatlantic Economic Partnership. Multilateral efforts include US involvement in the establishment of Codex Alimentarius food-safety standards and the various committees administering the World Trade Organization (WTO) agreements.

In 1996, the Clinton Administration published a report on the desired future for US food regulations. Among other things, this report encouraged efforts by US regulatory agencies to promote the international harmonization of food regulations. Of particular concern to agricultural trade policy is the presence of regulatory barriers to trade in US foodstuffs. For instance, the Foreign Agricultural Service (FAS) of the USDA attempts to identify trade barriers and to challenge their use through consultation in international forums, including the Technical Barriers to Trade (TBT) and Sanitary and Phytosanitary Standards (SPS) Committees. Public participation in trade-policy development is through the Advisory Committee on Trade Policy Negotiations. When trade concerns become trade tensions, the US Office of the Trade Representative (USTR) takes over. The USTR seeks to consult on the reduction or removal of trade barriers and acts as the relevant US agency in the event that trade disputes under international agreements arise. The USA has at times acted unilaterally to address perceived unfair trade practices of other countries, outside the scope of international agreements. According to Section 301 of the US Trade Act of 1974, the Department of Commerce can take action against the imports of foreign goods in response to 'unfair and discriminatory practices or breaches of trade agreements' by trading partners (Horton, 1997). This unilateral action is controversial and many trading partners feel that it undermines the rules-based trading regimes outlined in international trade agreements.

The US regulatory system and regulatory responses cannot be fully characterized without some mention of US product-liability laws. According to US law, legal liability rests with the product manufacturer and, in the event of a food- or environmental-safety incident, a consumer only has to show that the manufacturer was responsible for the product, not that the manufacturer was negligent. US product-liability laws are an important factor in the internalization of food- and environmental-safety risks by US manufacturers, because a safety risk quickly becomes a commercial crisis. Firms minimizing commercial risks in the USA are simultaneously minimizing safety risks.[3]

The US regulatory system may now be characterized according to the scientific- and social-rationality paradigms identified in Chapter 5 (Table 6.2). The general interpretation of the US regulatory approach is that it is overwhelmingly a scientific-rationality approach. Essentially, the US regulatory system is based on the belief that the new techniques of genetic engineering are an extension of traditional breeding techniques and biotechnology-based products are simply an extension of existing product classes. In this sense, US regulations have been built on the experience and expertise of, and the public confidence in, US regulators.

[3]See Spriggs and Isaac (2001) for a discussion of US product-liability laws in respect of the minimization of food-safety risks in the beef industry.

With respect to the first specific regulatory debates associated with GM crops, the US approach employs the scientific interpretation of the precautionary principle. There exists considerable confidence that over 25 years of scientific assessments on rDNA technology have shown that the risks from genetic modifications are not inherently different from the risks using more traditional techniques, such as somoclonal variations and mutagenesis. Given this view, the precautionary principle is used as a risk-assessment tool in order to determine hypothetical risks. Secondly, while APHIS and the EPA initially apply a process-based filter to GM crops in order to determine regulatory oversight, the US regulations are focused primarily on the product applications of genetic modification, not the technology *per se*. For instance, the FDA applies the substantial-equivalence principle to determine novelty. Thirdly, regulatory authority resides with vertical authorities according to traditional jurisdiction, where existing

Table 6.2. US regulatory-approach check-list.

	Scientific rationality	Social rationality
General regulatory issues		
Tradition	✓ Regulatory independence	
Belief	✓ Technological progress Enhances growth and development, leading to higher social regulations – a regulatory race to the top	
Type of risk	✓ Recognized ✓ Hypothetical	
Substantial equivalence	✓ Yes	
Science or other factors in risk assessment	✓ First regulatory hurdle	
Burden of proof	✓ Traditional: minimize Type I (rejected when safe)	
Risk tolerance	✓ Minimum	
Science or other factors in risk management	✓ Safety- or hazard-based Risk management is for risk reduction and prevention only	
Specific regulatory issues		
Precautionary principle	✓ Scientific interpretation	
Focus	✓ FDA: product-based, novelty	APHIS and EPA: initially process-based to determine regulatory oversight
Structure	✓ Vertical, existing structures	
Participation	✓ Narrow, technical experts ✓ Judicial decision-making	
Mandatory labelling	✓ Safety or hazard only	

statutes have been amended to deal with novelty. The rationale is that, given the product or application focus, there is not an appropriate horizontal structure, whereas existing health and safety laws provide immediate regulatory protection based on long-standing and trusted expertise. Fourthly, as agricultural biotechnology products are considered within the US regulatory structure, the decision-making remains largely narrow – limited to the traditional actors – and judicial, rather than consensual. Fifthly, the mandatory labelling policy in the USA supports a safety- or hazard-based justification only.

The US regulatory system may be assessed from both an economic perspective and a social perspective, which includes consumer concerns about food safety and environmental protection and moral and ethical concerns.

ECONOMIC PERSPECTIVE. From an economic perspective, the US system provides industry with predictable and timely decisions. A GM-crop regulatory submission requires three types of information (Kraus, 1998). First, the firm must submit information on the genetic characteristics of the host and recipient organisms, on the vector of transfer employed and on the resultant genetic characteristics of the GM crop. Secondly, information on the purpose, method of data collection and harvest procedures for the field trial must be submitted. Thirdly, information on the field-trial location and the supervision procedures must also be submitted. Once approval has been given on the particular GM-crop variety, the only regulatory responsibility is an annual field-trial report. This field-trial regulatory approach has been commercially friendly. For instance, APHIS received more than 920 requests for the environmental release of GM varieties between 1988 and 1998. Of these requests, 886 were approved. Also, between 1992 and 1999, APHIS received 65 petitions for deregulation, where 49 were approved in an average of 4 months, while 11 were voluntarily withdrawn and five were rejected. The EPA approved 34 proposals for the environmental release of biopesticides between 1987 and 1996. The FDA reviewed and approved the commercial release of 43 GM crops between 1994 and 1998 (APHIS, 1999).

The US approach creates commercial certainty and predictability. For instance, Fuchs (1993) has detailed the regulatory compliance efforts of Monsanto in seeking approval for the GM potato variety resistant to the Colorado potato beetle (CPB). The Monsanto strategy was twofold. The first strategy was to establish that the GM potato was substantially equivalent to conventional varieties and hence not subject to novelty regulations. The second strategy was to provide required data on the human-, animal- and environmental-safety performance of the GM potato. This is not to suggest that the US regulations are easy, but rather to suggest that, given the certainty and predictability, agricultural biotechnology firms know what regulatory hurdles they must overcome and the type of risk assessment information they must present in order to gain market approval. Therefore, US agricultural biotechnology firms favour the predictability and stability of the US regulatory approach, because it clarifies the information requirements that firms must meet. This is reflected in high rates of application approvals with few regulatory hold-ups. It has been argued that 'the US plant biotechnology industry has been positively affected by domestic regulation and finds itself at a comparative regulatory advantage to its European counterpart' (Kraus, 1998). In short, the US scientific-rationality approach to regulating GM crops benefits the economic interests by ensuring a predictable framework encouraging technological progress.

SOCIAL PERSPECTIVE. From the social perspective, it has been argued that the US regulatory agencies, such as the FDA and the EPA, are trusted by the public to protect both the food supply and the environment (Hoban, 1997). It has also been argued that 'it is not an overstatement to suggest that the level of scientific research and testing involved in biotechnology [agricultural] is at least as good as that available in the health care field' (Balk, 1993). Public trust in US medical regulations is high and capturing that level of trust was an important concern of US regulatory development, where it was viewed that attempts to develop a new set of horizontal regulatory rules or a new regulatory agency might have undermined the public trust. Kraus (1998) argued that the decision to keep regulatory oversight within the traditional vertical jurisdictions is the reason for increased consumer confidence and public acceptance in the USA.

The US approach does, however, have significant shortcomings in dealing with the social dimensions of agricultural biotechnology. For instance, the important role of APHIS may be construed as a conflict of interest, since the USDA is playing the dual role of promoting the agricultural sector while at the same time regulating it. This raises concerns about regulatory capture and who the clients of the USDA are – industry or consumers. In fact, other countries such as Canada and the UK, have taken steps to separate the promotion and regulation activities into different regulatory agencies.[4] Also, limited participation in the regulatory decision-making process may increase fears that agricultural biotechnology is being approved without adequate public consultation or consent.

If GM foods become as politicized in the USA as they currently are in the UK, the US regulatory system will be under considerable

[4]In Canada, Agriculture and Agri-Food Canada (AAFC) promotes, while the new Canadian Food Inspection Agency (CFIA) and Health Canada (HC) regulate, and, in the UK, the Ministry of Agriculture, Fisheries and Food (MAFF) promotes, while the new Food Standards Agency (FSA) will regulate.

pressure to consider the social dimensions of agricultural biotechnology, and yet it may be poorly equipped to do so. The ability to transform the vertical regulatory agencies in a manner sufficient for the interest groups, without changing the underlying regulatory principles on which the system is based, appears very limited. This threat has been at the heart of proactive, coordinated efforts in the USA by government and industry to stop any 'social contagion' resulting from European consumer reactions (*Economist*, 1999a). The strategy is to investigate scientific, economic and social concerns through broad participation in advisory committees supported by the National Economic Council and the USDA. In addition, the Grocery Manufacturers of America, an industry association, is determined to prevent consumer concerns from forcing product reformulation on its members, as happened in the UK. In other words, industry is working in coordination with government to protect against potential regulatory instability, which could have serious commercial implications.

6.3 Agricultural Biotechnology in Canada

Biotechnology in Canada is considered to be vital to future growth prospects. Although global-capacity leadership is enjoyed in several specific areas, such as the genetic modification of canola, overall, biotechnology capacity in Canada tends to lag behind that of the USA as well as those of Japan and several European countries.

6.3.1 Canadian promotion of agricultural biotechnology

The Canadian federal government, along with some provincial governments, has been very proactive in the development and commercial application of modern biotechnology to Canadian products. In 1983, the National Biotechnology Strategy was established, committing approximately C$13 million year^{-1} to biotechnology research through National Centres of Research Excellence. It has been estimated that government promotion has averaged C$250 million year^{-1} or over C$1500 million for the decade 1987–1997 (CIB, 1997).

Unlike that in the USA, public policy in Canada directly targets the deepening and widening of the Canadian biotechnology capacity. According to the Canadian biotechnology strategy, the application of biotechnology in the production of agricultural, industrial and medical goods is held to be a potential cornerstone of the economy (Industry Canada, 1997). Further, the Canadian Institute of Biotechnology (CIB) argues that 'biotechnology is an important driver of economic growth in Canada. It has become an important tool that, when used appropriately, can increase the competitiveness of all sectors from agriculture to health care' (CIB, 1997).

While Canada does not hold a commercialization lead in agricultural biotechnologies over the USA, it does have a lead over the European Union (EU), resulting in a North American lead in the commercialization of GM crops (Phillips and Khachatourians, 2001).

6.3.2 The Canadian regulatory system

In Canada, the traditional regulatory role of the state has been primarily influenced by the USA, due to the economic importance of the US market to Canadian products, although there has been some European influence, such that the Canadian approach blends the regulatory-independence approach of the USA, with the European regulatory-accountability approach. For instance, as in the USA, the regulatory-independence approach prescribes government regulatory intervention only in reaction to market failure. Similarly, Canadian regulatory decision-making is traditionally done by regulators, limiting the influence of populist politics and day-to-day public interests on the regulatory system. However, unlike in the USA, Canadian regulators also examine the 'effectiveness' of regulations. This is more closely associated with the European approach, employing the fourth regulatory hurdle (i.e. safety, health, quality and effectiveness or social benefit). With respect to modern biotechnology, it will be argued that the Canadian regulatory approach has been pro-competitive, focused on ensuring technological progress.

The Canadian regulatory system for GM crops was driven endogenously by the development of GM canola/rape-seed varieties and exogenously by the rapid pace of R&D and commercialization of GM crops in the USA. With respect to the endogenous drivers for regulatory development, it has been argued that the regulatory system was developed essentially to support the canola industry and tended to take its lead from the industry (Phillips and Khachatourians, 2001). In this sense, the traditional regulatory objective has been one of technological progress, with a focus on product applications consistent with the scientific-rationality paradigm, and the entire Canadian regulatory structure flows consistently from this initial trajectory.

Pressure to formalize the Canadian regulatory system also came from the imminent commercial approval of GM varieties of maize, soybeans, tomatoes, cotton and squash in the USA. Therefore, it was crucial for Canada to develop an acceptable level of regulatory oversight for the presence of novel plants in the Canadian food supply. The result was the development of the *Guidelines for the Assessment of Novel Foods*, which is essentially the foundation of the Canadian regulatory approach (Health Canada, 1994).

Functionally, the regulatory system in Canada is established as follows. Health Canada (HC) is responsible for the scientific risk assessments required to determine food-safety policies and regulations. Similarly, Environment Canada (EC) is responsible for the environmental-impact assessments to determine environmental-safety regulations. Consistent with the Redbook on risk analysis, the risk-assessment activities of these two agencies are institutionally separated from the risk-management activities of the Canadian Food Inspection Agency (CFIA), which is responsible for implementing, monitoring for compliance and preventing contravention of the policies developed by HC and EC. The leading regulatory division of the CFIA is the Plant Products Division, which is advised by the Advisory Committee on Plant Biotechnology Regulation. The objective of the Advisory Committee is to assist the Plant Products Division in identifying and developing principles and protocols for the regulation of GM plants.

To receive commercial approval for a GM variety in Canada, companies must supply risk-assessment data and information: to HC, in order to satisfy the Novel Foods Guidelines; to Agriculture Canada, for animal-feed approval and variety registration; to the CFIA, to satisfy the field-testing regulations; and to EC to satisfy the environmental-impact assessment regulations. Given the dominance of the scientific-rationality paradigm in Canada, the risk information may be from hypothetical risk assessments in order to determine either the substantial equivalence or the novelty of the GM crops.

In late 1999, HC announced amendments to the Canadian Food and Drugs Law, established to clarify the approval process for novel foods. The amendments followed 7 years of consultations with all stakeholders (*World Food Law Monthly*, 1999b). There were three main amendments. First, a detailed definition of novel foods was established:

(a) a substance, including a microorganism, that does not have a history of safe use as a food;
(b) a food that has been manufactured, prepared, preserved or packaged by a process that:
    (i) has not been previously applied to that food,
    (ii) causes the food to undergo a major change; and
(c) a food that is derived from a plant, animal or microorganism that has been genetically modified such that:
    (i) the plant, animal or microorganism exhibits characteristics that were not previously observed in that plant, animal or microorganism;
    (ii) the plant, animal or microorganism no longer exhibits characteristics that were previously observed;
    (iii) one or more characteristics no longer fall within the anticipated range for that plant, animal or microorganism.

The definition also specified that there are three types of variations that place the foodstuff outside the anticipated range: variations in the composition, structure or nutritional quality outside the conventional range; variations in the manner in which the food is metabolized in the body; and variations in the microbiological safety, the chemical safety or the safe use of the food relative to conventional products.

Secondly, given the clarifications of the definition of novel, the amendments detailed procedures for assessing the safety of novel foods. The substantial-equivalence principle remains a gateway principle for determining if a GM crop is to be regulated as a conventional crop or as a novel product. A notable concept here is that of 'prior safe use as a food', which allows HC to approve the notification of a novel food that has been accepted by certain other countries. In this sense, this concept represents a form of regulatory mutual recognition.

Thirdly, it established mandatory premarket notification for the commercialization of novel foods. HC had considered premarket approval but found that this would add little to the safety while adding extensively to the cost of commercialization. Premarket notification requires 45 days' prior notice to HC before sale or advertisement of any novel food. HC must give its response on the acceptability of the novel food within the 45 days. Additionally, HC can suspend the sale or advertisement if additional information is required.

There is a horizontal regulatory safety net in Canada, EC's Canadian Environmental Protection Act (CEPA). Novel products that do not fall under the regulatory oversight of either the CFIA or HC are under the requirements for notification of new substances of the CEPA. The federal regulatory system is identified in Table 6.3.

Table 6.3. Canadian agricultural biotechnology regulations.

Agency	Regulatory oversight	
Vertical legislation		
Health Canada	Novel-foods guidelines Responsible for scientific risk assessments for development of food-safety policies	Novelty-based
Environment Canada	Responsible for scientific risk assessments of microbial/plant pesticides, new uses of existing pesticides, novel microorganisms to develop environmental-safety policies	Novelty-based
Canadian Food Inspection Agency	Responsible for risk management, implementing, monitoring and enforcing HC and EC policies	Novelty-based
Horizontal legislation		
Canadian Environmental Protection Act (CEPA)	Requires the relevant regulatory authority to conduct an assessment of the environmental impact of all activities that are not already assessed under vertical regulation	

An important aspect of Canadian regulatory oversight for GM crops is that, unlike US regulations, there is scope for Canadian regulators to base decisions upon 'effectiveness' or social benefit. As will be shown, this is similar to the European fourth hurdle of regulatory oversight (safety, health, quality and effectiveness or socio-economic benefit). As was previously discussed, US regulatory decisions tend to adopt the position of letting the market decide a product's social benefit, provided the product is safe or of sufficient quality and exhibits its claimed attributes.

The Canadian policy on the mandatory labelling of products made from GM crops consistently flows from its scientific-rationality paradigm focused on novelty. As the Canadian regulatory system is product-based, not process-based, the labelling policy does not support process-based labelling for the consumer's right-to-know that GM techniques have been used. Only if the product has been determined to be novel is there a consumers' right-to-know justification for a mandatory labelling strategy. An added dimension to the Canadian labelling policy is that Canada is currently the chair of the Codex Committee on Food Labelling (discussed in Chapters 3 and 5).

Canadian efforts at international regulatory integration are led by the CFIA's Biotechnology Strategies and Coordination Office, which coordinates the Canadian regulatory policies with those developed at international forums, such as the WTO and the international negotiations for the Cartagena Protocol on Biosafety. It also provides advice to trade policy-makers and trade negotiators in both Agriculture and Agri-Food Canada (AAFC) and the Department of Foreign Affairs and International Trade.

Similarly to the US regulatory system, the Canadian system may be characterized according to the regulatory principles of the scientific- and social-rationality paradigms (Table 6.4). The general interpretation of the Canadian approach is that it is largely congruent with the scientific-rationality paradigm. Indeed, it is much more dedicated to the novelty basis of regulations than the USA. There is, however, an important deviation regarding the debates over science vs. other legitimate factors in risk assessment and risk management. In the risk-assessment procedure, the Canadian approach permits food products to be assessed according to all four regulatory hurdles. In the risk-management procedure, the Canadian approach permits other legitimate factors, such as socio-economic factors, in risk-management decisions, rather than just a narrow focus on risk reduction and prevention.

With respect to the specific regulatory principles associated with GM crops, the regulatory approach employs the scientific interpretation of the precautionary principle. Secondly, the system is novel-product-based, where substantial equivalence is assumed for all non-novel GM products. Thirdly, regulatory authority is associated with vertical jurisdictions. Fourthly, participation in the regulatory decision-making process is narrow and judicious, in that it is limited to traditional actors and experts. Finally, the mandatory labelling policy does not support GM labelling for the consumers' right to know. The scientific-rationality trajectory of the Canadian regulatory approach may be assessed from both an economic and a social perspective.

ECONOMIC PERSPECTIVE. From an economic perspective, the Canadian system has operated relatively efficiently, approving by 1997 more than 3800 field trials, 29 plants with novel traits for feed and three novel foods. It has been able to achieve this flow of regulatory decisions by considering substantial equivalence and by limiting

Table 6.4. Canadian regulatory-approach check-list.

	Scientific rationality	Social rationality
General regulatory issues		
Tradition	✓ Regulatory independence	
Belief	✓ Technological progress Enhances growth and development, leading to higher social regulations – a regulatory race to the top	
Type of risk	✓ Recognized ✓ Hypothetical	
Substantial equivalence	✓ Yes	
Science or other factors in risk assessment		✓ Risk assessments may be based on all four regulatory hurdles
Burden of proof	✓ Traditional: minimize Type I (rejected when safe)	
Risk tolerance	✓ Minimum	
Science or other factors in risk management		✓ Risk management may consider other legitimate factors beyond risk prevention and reduction: socially responsive
Specific regulatory issues		
Precautionary principle	✓ Scientific interpretation	
Focus	✓ Product-based, novelty	
Structure	✓ Vertical, existing structures	
Participation	✓ Narrow, technical experts ✓ Judicial decision-making	
Mandatory labelling	✓ Substantially equivalent: safety or hazard only ✓ Novel: consumers' right to know	

participation in the decision-making process to the proponent and the regulators. The commercial effectiveness of Canadian regulations on GM crops is favoured by the biotechnology industry. A 1994 survey revealed that 'improving the regulatory approval process' was considered the most important task for government by the biotechnology industry, and support for the industry has influenced regulatory development (Ernst and Young, 1995).

SOCIAL PERSPECTIVE. Citizens and consumer and environmental organizations, along with provincial governments, while allowed relatively free access to applications and specific decision documents through the CFIA website, have limited say in the science-based system. This has been somewhat controversial but has not created any significant backlash by either the major lobby groups or the general public in the magnitude of the European reaction.

From a social perspective, the Canadian system for food safety in general has been proactive in establishing an institutional framework designed to build consumer confidence. In 1997, all food inspection and assessment functions were transferred from AAFC to the new CFIA. Essentially, governmental promotion of the agriculture sector was separated from regulatory activities associated with agri-food and environmental-protection issues. By all accounts, this institutional restructuring has further enhanced consumer confidence in the government's regulatory role. Specifically, regulatory oversight for agricultural biotechnology is shared between HC, which assesses novel GM foods, and the CFIA and EC, which address specific consumer and environmental-health and safety issues associated with GM agricultural products.

While this approach has worked well up to this point, similarly to the US regulatory system, the Canadian system may have difficulty in incorporating social dimensions into the regulatory process in the event that the use of GM materials in the both the food supply and the environment becomes a highly politicized issue. Citizens and consumer and environmental groups, as well as provincial departments and agencies, who have relatively little access to the decision-making process, may demand both greater transparency and participation.

In a proactive effort to address the social perspective on biotechnology, several government initiatives are worth noting. First, the Canadian Biotechnology Advisory Committee (CBAC) has been established as the federal government's conduit for the two-way flow of information between regulators and the general public. A subcommittee of CBAC has a remit to focus on consumer acceptance of the regulations associated with GM crops and foods derived from GM crops. Its report was expected in July 2001. Secondly, in late 1999 the Federal Government announced the creation of a Scientific Advisory

Committee under the Royal Society of Canada to examine what type of agricultural biotechnology products are coming, what the anticipated short- and long-term health implications are, what the international developments are in this area and, consequently, what type of scientific regulatory capacity is required in Canada in order to ensure the safety of agricultural biotechnology products. The Royal Society Report, released in early 2001, has largely been interpreted as a severe criticism of the scientific-rationality approach to regulating GM crops. In response, the Canadian federal government noted that, while the Royal Society report does raise some specific issues that need to be addressed as novel-output trait and bioengineered products come down the pipeline, the interpretation of the existing structure was, in many instances, wrong. Thirdly, under the auspices of the Canada Genome project, intended to fund major scientific research initiatives in the fields of biotechnology, there is a component to support social-science research about the technology simultaneously. This is known as the genome, ethics, law and society (GELS) component. It is expected that combining natural science with GELS research will continue to be the focus of the Canadian federal government.

6.4 North American Regulatory Integration

Given the significant linkages between Canadian and US agriculture and especially the respective research efforts, regulatory integration was necessary to provide consistent regulatory oversight for industry promotion and consumer and environmental protection.

The two leading regulatory agencies – the USDA's APHIS and the CFIA – have worked to streamline the approvals of products in the two countries. In the past, the two agencies have undertaken simultaneous reviews of GM plants prior to their commercialization and have shared data and observations informally. In 1998, APHIS, the CFIA and HC regulatory officials met to compare and harmonize, where possible, risk-assessment procedures, such as the molecular genetic-characterization components of the regulatory review process for GM plants – that is, they examined opportunities to streamline the science-based and rules-based approach to risk assessment.

The Canadian–US integration of agricultural biotechnology regulations is congruent with agricultural trade-liberalization commitments made under the original Canada–US Trade Agreement (CUSTA) and continued in the NAFTA. Both agreements are free-trade agreements based on sovereignty and regulatory competition, with no goal of deep integration. The objective, in general, is to engage in a long-term regulatory integration but within a traditional trade framework. For instance, under the CUSTA, the USA and Canada established

Technical Working Groups (TWGs) to deal with specific sectoral trade issues. The most important TWG for the issues dealing with GM crops is that on seeds. Under the NAFTA the TWGs were trilateralized and two permanent committees were established to ensure continual dialogue on issues of regulatory barriers to trade: the Committee on Standards Related Measures and the Committee on Sanitary and Phytosanitary Measures. Further, a 1995 NAFTA Memorandum of Cooperation called for both the regular reporting of food standards to improve coordination efforts and to coordinate a 'NAFTA' position on food standards in international forums, such as Codex and the WTO.

This level of regulatory integration between the two countries is made possible because the regulatory approaches both reflect the scientific-rationality approach to regulating GM crops. Further, the regulatory integration ensures that there are few insurmountable social regulatory barriers to trade.

There are, however, important limits to the type of regulatory integration pursued under the NAFTA. Most importantly, the regulatory integration remains subject to the traditional trade principle of regulatory sovereignty. That is, the regulatory integration would not be likely to work in the event that one jurisdiction employed a scientific-rationality approach while the other jurisdiction employed the social-rationality approach.

6.5 Conclusions

The North American approach to regulating GM crops is summarized in Table 6.5. In general, it reflects the scientific-rationality paradigm encouraging technological progress, driven initially by scientists many years before commercialization issues were raised. It is sup-

Table 6.5. Characterization of North American regulatory approach.

Regulatory Approach	
Tradition	Regulatory independence
Paradigm	Scientific rationality
Specific regulatory debates	
Precautionary principle	Scientific interpretation: risk-assessment tool
Focus	Product- or application-based, novelty focus
Structure	Vertical, existing regulatory agencies
Decision-making	Narrow participation of experts Judicious
Mandatory labelling	Substantially equivalent: hazard- or safety-based Novel: consumers' right to know
Integration	Economic integration according to regulatory competition

ported and influenced by economic interests and, as a result, it is conducive to the North American commercialization lead in agricultural biotechnologies.

While being commercially friendly, this approach does have some significant drawbacks in its social responsiveness. It is unlikely that the North American approach can shift enough towards the social-rationality paradigm to meet social interests without abandoning the scientific regulatory principles. In other words, should a crisis of confidence emerge in North America of the magnitude of that in the UK, the North American approach might have to make some rapid amendments in favour of social-rationality principles.

Regulatory integration dealing with social regulatory barriers is made easier when two jurisdictions have adopted the same regulatory principles, as with Canada and the USA. The difficulty is, of course, integrating regulatory approaches that have adopted divergent paradigms.

7 European Regulatory Approach

7.1 Introduction

Similarly to the North American regulatory approach, the European approach must strike a balance between technological progress, in order to remain competitive in the frontier technology, and technological precaution, in order to assuage consumer concerns. The objective of this chapter is to examine both the promotion and regulation of genetically modified (GM) crops at the European Union (EU) level and within the UK.

Prior to this, it is crucial to identify the European perception of the agriculture sector, as this provides important context or background to European efforts to both promote and regulate GM crops. As previously mentioned, in North America, agriculture is perceived to be an industry. This is very different from the general perception of the agricultural sector in Europe, which is generally perceived (especially in Western Europe) to be more than just an industry. It fulfils a 'multifunctional' role, including supporting the rural way of life, preserving the culture and heritage of the countryside, ensuring the welfare of animals (both livestock and wildlife) and protecting the environment. Accordingly, agriculture is a special kind of endeavour, forming the last line of defence between the country and urbanization (Haniotis, 2000; Kalaitzandonakes, 2000). In addition, European consumers have faced several severe food-safety crises that have undermined the public trust in regulators and scientists (Consumers' Association, 1997; Grove-White *et al.*, 1997; Sheppard, 1997; National Consumers Council, 1998; Gaskell, 2000; Sylvander and Le Floc'h-Wadel, 2000). The implication of this for credence goods, such as GM crops, is that there is little trust among consumers in the traditional sources of risk information. Combining the multifunctionality role with the lack of public trust results in a European agricultural sector under considerable scrutiny as a well-politicized issue.

7.2 Agricultural Biotechnology in the European Union

There is considerable capacity in agricultural biotechnologies in Europe. For instance, firms on the frontier of GM-crop development include Aventis and Syngenta. The European promotion and regulation of GM crops must navigate the relationship between the EU level and the approaches and initiatives of the various member states. The competing forces of regulatory centralization and subsidiarity exist not only between the EU level and trading partners, but within the EU between European services and member states.

7.2.1 EU promotion of agricultural biotechnology

Developing a competitive posture in biotechnology has been a strategic priority in Europe for nearly two decades. While the EU is behind North America in the commercialization of GM crops (James, 1999; Phillips and Khachatourians, 2001), significant research efforts, both public and private, are under way to close this gap.

Originally, DG XII (Science, Research and Development) led the push, although there have been several initiatives, including:[1]

1. Biotechnology Framework Programme (BFP), 1983 – promotion and marketing.
2. Biotechnology Engineering Programme (BEP), 1982–1986.
3. Biotechnology Action Programme (BAP), 1986–1989.
4. Biotechnology Research and Innovation for Development and Growth (BRIDGE).
5. Euro Collaborative Linkage of Agriculture and Industry through Research (ECLAIR) – ECU80 million to develop new plants for industrial uses.
6. Food-Linked Agro-Industrial Research (FLAIR) – transnational food-safety, quality and competitiveness projects, creating value for agricultural biotechnology products.
7. European Commission's Innovation Programme – a supply-driven initiative to support the commercialization of advanced technologies.

Along with these promotion programmes, the EU regulatory approach to biotechnology was also reviewed on several occasions in

[1]From Sorj *et al.*, 1989; Haerlin, 1990; Theodorakopoulou and Kalaitzandonakes, 1999.

order to ensure that it did not unacceptably hinder technological progress. In 1983, DG XII examined the competitive position of European biotechnology relative to the respective positions in the USA and Japan. The Commission's report, entitled *Biotechnology in the Community* (European Commission, 1983), sought to develop a comprehensive biotechnology strategy for the European Economic Community (EEC) that would improve its global competitiveness. In 1991, the Commission's Biotechnology Coordinating Committee presented a paper to the European Council and the European Parliament, entitled Promoting a *Competitive Environment for Industrial Activities Based on Biotechnology Within the Community*, outlining regulatory reform required to improve competitiveness. In 1994, the importance of biotechnology in the EU was again highlighted by the White Paper on Growth, Competitiveness and Employment, which stated that 'the growth rate of industries based on modern biotechnologies is expected to be substantial' (European Commission, 1994a). In each of these policy documents, the regulatory structure was explicitly considered to be a key to the promotion of biotechnology in the EU. As other nations continue to invest both publicly and privately in the research, development and commercialization of new agricultural biotechnology products, an important competitiveness imperative is created to build the EU capacity, and a regulatory system encouraging technological progress is viewed as one way to do this. It has been argued, in fact, that, if the EU does not develop such a regulatory approach and instead continues with its restrictive regulatory policies, the result will be a global allocation of biotechnology capacity where North America will be the dominant region for research and development into biotechnology, developing countries will become the producers of resultant GM crops and Europe will become the dominant producer of labour-intensive sunset technologies (Weatherspoon *et al.*, 1999). Such a result seems unacceptable from the point of view of the 1994 White Paper.

Joly and Lemarie (1998) have argued that Europe is already at a competitive disadvantage in GM-crop technologies targeted at production-improved varieties and, due to cumulative innovation effects, it is unlikely to catch up. From a competitiveness position, the solution is more direct promotion of GM-crop development focused on output-improved varieties, bioengineered products and genomics. Indeed, a recent industrial promotion initiative in France has the mandate of closing the technology gap. The GenoPlante Initiative targets the development of GM varieties of rice, maize, wheat and rape, which are important crops in French agricultural production. Through this initiative, public research (INRA, CIRAD, ORSTOM, Centre National de la Recherche Scientifique (CNRS)) is coordinated with private research efforts (LimaGrain/BioGemma and Rhone-Poulenc). If successful, it can be expected that this type of initiative will also be adopted at the EU level (Joly and Lemarie, 1998).

Along with France, other member states also have biotechnology promotion initiatives (*Financial Times*, 1999c). For instance, in 1995, the UK launched the 'Biotech Means Business' initiative, followed up in 1998 by the £13 million BioWise programme designed to encourage industry to use biotechnologies. It has been estimated that, up to 1997, the UK had spent £176 million through the Biotechnology and Biological Sciences Research Council (Sheppard, 1997). The dominant biotechnology programme in Germany has been BioRegio, an initiative that provides matching Federal–Länder funds for start-up biotechnology companies. However, in late 1999, Germany announced a new initiative, the 'BioProfile Programme', which aims to spend DM100 million on the promotion of German biotechnology to establish Germany as the European leader in the field (*AgraFood Biotech*, 1999b).

Capacity in GM-crop development and commercialization specifically – as well as other agricultural biotechnologies in general – takes on an even more important competitiveness aspect in the context of the Common Agricultural Policy (CAP). Under significant political pressure, the CAP faces restructuring that would involve the retreat of domestic farm-support payments. Agricultural biotechnology can potentially fill the gaps by providing scientific solutions to agricultural production problems of variability in quantity and quality.

The EU promotion of GM-crop development has, however, come under pressure from critics. Mindful of the implications of a loss in consumer confidence, in 2000 the European Commission announced cuts in explicit research funding for the support of GM-crop development, because of consumer concern. Spending will continue in support of basic biotechnology techniques and procedures.

7.2.2 The EU regulatory system

It has been argued that the European regulatory approach reflects an accountability tradition where government regulatory intervention is dominated by concerns over the democratic accountability of the discretionary decision-making power of regulators (Majone, 1994). In achieving accountability, final regulatory decision-making resides with elected public officials, as they are viewed to be the most representative of the public interest. Yet, as a consequence, regulatory development and administration are very much subject to the day-to-day public interests dominating the concerns of elected officials.

As mentioned in Chapter 5, in 1985, the European Commission published its White Paper on the Single European Market, proposing to remove the politics from regulatory convergence among member states. According to the 'new approach' to regulatory development and integration, both food-safety and environmental-protection regulations pursue a dual objective: to ensure the free movement of products in the internal market while protecting consumers and the environment (Woolcock, 1996). Yet there are important differences between food-safety and environmental-protection regulations. The legal basis of EU food-safety regulations is derived from Article 100a of the Treaty Establishing the European Communities, in which facilitation of the internal market is the primary objective. On the other hand, the legal basis of environmental-protection regulations tends to be Article 130r Title XVI Environment of the Treaty Establishing the European Communities, in which protection of the environment is paramount, not facilitating the internal market. According to this article, member states may unilaterally impose regulations that bar market access in order to protect the domestic environment. Hence, while food-safety regulations are compelled to be market-sensitive, environmental-protection regulations are not. It will be argued that this difference in the legal basis encourages a shift in European GM-crop regulations from food-safety justifications to environmental-protection justifications, which are not subject to the same requirements of facilitating the common market.

Given this regulatory context, the objective now is to identify the development of GM-crop regulations in the EU. It is important to note that there were, and continue to be, divergent views between member states and the EU services and also divergent views within the various services of the EU on the benefits and risks of agricultural biotechnology, generating important debates regarding appropriate regulatory principles. For instance, in respect of divergent views of member states, there have been several cases where members have imposed unilateral moratoriums on the field-trial plantings of GM crops. While these instances will be examined in greater detail below, it is important to note that membership in the EU has meant that members lose the sovereign right to impose unilateral regulatory actions, and the legality of these moratoriums is in doubt. Also, among the DGs of the European Commission, there are levels of support for the scientific- and social-rationality paradigms for the risk-analysis framework.

The first proposed regulatory policy dealing with the research, development and commercialization of biotechnology was developed by DG XII (Science, Research and Development) in 1978. This proposal was for a binding Council Directive that would require notification to and authorization from the competent national authority for all biotechnology research. The competent authority would then advise DG XII on approvals, in order to build up a base of biotechnological knowledge about risks, problems and experiences. This proposed directive was notable for three reasons. First, it employed the scientific interpretation of the precautionary principle, which was also recommended in the 1976 US National Institutes of Health (NIH) guidelines. According to this interpretation, controlled laboratory research was permitted, even though the risks were uncertain, because such research would produce risk information, leading to a better understanding of the risks. Secondly, the proposed directive was horizontal in nature, applying to the process and production methods of all biotechnology research, regardless of the proposed or intended end-use. The reason for horizontal guidelines was to create an essential minimum requirement for research activities across all member states. Yet this proposal was in conflict with the 1976 NIH guidelines, which suggested that there were no significant risks associated with the use of biotechnology; rather, the risk was linked to the proposed end use. The NIH recommended product-based regulations, while the proposed EU directive was for technology-based regulations. Thirdly, the proposed directive supported regulatory subsidiarity, allowing competent authorities in each member state to approve or reject research initiatives, rather than centralizing decision-making at the EEC level.

By 1980, the directive proposal was replaced by a proposal for a non-binding recommendation, adopted by the Council in June 1982 as the Council Recommendation Concerning the Registration of Work Involving Recombinant Deoxyribonucleic Acid. The proposed directive was softened to a non-binding recommendation in reaction to the relaxation of the 1976 guidelines by the NIH in the USA. Essentially, the scientists active in the nascent research on biotechnology projects realized that previous, precautious assumptions about the hypothetical risks from genetic modification were exaggerated, and guidelines and recommendations were relaxed.

In 1983, DG XII published the report *Biotechnology in the Community*, which, from a regulatory perspective, was critical for two reasons. First, the report deviated from the 1982 Council recommendation by not supporting a horizontal, process-based regulatory policy for biotechnology. Instead, it supported the product-based regulatory approach that appeared favourable to technological progress and biotechnology competitiveness in the USA: that is, the report urged the European Community (EC) to adopt a regulatory structure more congruent with the emerging US approach. Secondly, anticipating the public response to modern biotechnology, it sought 'to ensure regulatory provisions to maintain national standards of public safety; and to this end, to monitor the social dimensions of biotechnology' (European Commission, 1983). Including the vague reference to the social dimensions of biotechnology essentially pried open the regulatory decision-making structure of the EU. According to the report, participation in the previously narrow and judicial structure should be expanded to include those actors who represented the social dimensions – that is, the expert scientific and technocratic participants were joined in the decision-making process by social interests, such as consumer and environmental organizations, who claimed to represent the social dimensions of biotechnology in the EU.

In 1986, the European Commission submitted a communication to the European Council entitled *A Community Framework for the Regulation of Biotechnology* (European Commission, 1986). This communication was the Commission's response to a 1986 Organization for Economic Cooperation and Development (OECD) recommendation on how to regulate biotechnology appropriately. Consistent with the OECD recommendation, the European Commission's response called for a 'clear, rational and evolving basis for biotechnology regulation' and for regulatory harmonization of the EC regulations with those of other countries. At this point, the North American and European regulatory approaches to biotechnology shared regulatory principles and were on a similar trajectory. For instance, the focus on product-based regulations, the scientific interpretation of the precautionary principle and the regulatory structure in existing vertical agencies were supported by the OECD and other international organizations. Indeed, international harmonization appeared to be a real possibility. This was not to last. Internal European conflicts within and among the Commission, the Parliament and various member states significantly challenged regulatory principles such as a product-based, novelty focus and the scientific interpretation of the precautionary principle.

An important challenge to the regulatory trajectory proposed in the European Commission's 1986 communication was the European Parliament (EP)'s Viehoff Report, released in 1987, which recommended regulations for the environmental release of genetically modified organisms (European Parliament, 1987). This report was an initiative of the European Parliament's Committee on Energy, Research and Technology, and the rapporteur was Phili Viehoff. Six Parliamentary Committees were involved: Environment and Consumer Protection; Economic Affairs and Industrial Policy; Agriculture and Food; Energy, Research and Technology; Social Affairs and Employment; Legal Affairs and Citizens' Rights. The report recommended that the 'Commission give priority to studying the problems posed by the potential release into the environment of genetically engineered natural microorganisms' and demanded a moratorium on such releases 'until binding Community safety directives have been drawn up'. The report also called for the harmonization of member states' procedures for risk assessment, in order to establish common environmental-protection measures.

The Viehoff Report was a crucial turning-point in the development of EU regulations for GM crops. First, the range of parliamentary committees involved, indicative of the broad range of European concerns, established a precedent for what was meant by the social dimensions of biotechnology. Secondly, the report's recommendations explicitly suggested that technology-based, binding Community directives were necessary, rather than a patchwork of national regulatory approaches. Community regulatory harmonization and centralized decision-making – as opposed to a subsidiarity approach – were viewed as the only appropriate way to deal with the transnational uncertainties and risks associated with the environmental release of genetically modified organisms. In fact, this was a direct attempt to prevent unilateral member state action and to establish a coordinated Community-wide regulatory approach. Thirdly, the report focused directives on Community environmental safety and health, rather than on facilitating the single market. This was a crucial development, as will be shown, because the report provided a foundation for DG XI (Environment) to argue that regulatory precaution in biotechnology regulations must be according to the social interpretation of the precautionary principle codified in the legal basis of Article 130r of the Treaty Establishing the European Communities. Until these regulatory changes were implemented and binding, harmonized and horizontal directives were in place, the report called for a moratorium on research into and commercialization of products derived from biotechnology. Therefore, the Viehoff Report departed significantly from the scientific-rationality regulatory trajectory of product-based regulations, employing the scientific interpretation of the precautionary principle. Indeed, this report was essentially a strong move towards the social-rationality regulatory tajectory.

Not all regulatory debates were between the European Commission and the EP. Despite the 1986 publication of the Commission communications, there was considerable intra-Commission conflict, due both to genuine differences in regulatory ideology and philosophy and to territorial disputes between vertical regulatory jurisdictions (Patterson, 2000). In terms of ideological and philosophical differences, DG III (Industry), DG VI (Agriculture) and DG XII (Science, Research and Technology) all basically supported the scientific-rationality paradigm and regulatory principles adopted in the 1976 NIH guidelines and the 1986 OECD recommendations and supported by Canada and the USA. The traditional 'clients' of these directorates were mostly industry and hence commercial objectives remained influential. This is not to suggest, however, that these directorates were not interested in protecting human and environmental health and safety. They simply perceived that the potential risks or hazards were from products, not from technology, and that the regulatory focus should be on the novelty of the product or application.

On the other hand, the ideology and philosophy of DG XI (Environment), with support from DG XXIV (Consumer), favoured a social-rationality approach to regulating risks, supporting new horizontal directives cutting across the various mandates and jurisdictions of vertical directorates. Further, wide actor participation was supported in regulatory decision-making, in order to preserve the focus on the social dimensions of biotechnology. The social interpretation of the precautionary principle was to be employed to ensure the highest level of consumer and environmental protection in the face of speculative risks. This regulatory philosophy was congruent with the Viehoff Report and hence enjoyed significant support from the EP. Since the 'client base' of these two directorates included consumer and environmental organizations, an important objective of the regulatory focus was on assuaging public concerns and fear by creating the impression that biotechnology was well controlled both in the food supply and in the environment (Patterson, 2000). In essence, the regulatory focus was on addressing consumer perceptions of risk, regardless of the scientific rationality of the risk perceptions or the hypothetical risks demonstrated through scientific risk-assessment procedures.

In terms of the territorial disputes across vertical regulatory jurisdictions, it has been argued 'no DG wanted to cede autonomy and regulatory authority to another DG in an area which they believed fell squarely into their policy domain' (Patterson, 2000). Yet, since biotechnology is a process facilitating product development across a wide spectrum of product categories, jurisdictional overlap was inevitable and unavoidable. Several intra-Commission initiatives were established to deal with jurisdictional overlap and to facilitate improved regulatory coordination within the Commission. In response to the 1983 Commission report on European competitiveness in biotechnology, the Biotechnology Steering Committee (BSC) was established in 1984 (Cantley, 1995). While four directorates were initially involved (DG III, DG VI, DG XII and DG XIII (Information, Market and Innovation)), participation was open to all DGs where concerned. Administratively, DG XII chaired the BSC, while the directorate's Concertation Unit for Biotechnology in Europe (CUBE) was the secretariat. DG XI became involved in the BSC when potential environmental risks were identified and discussed. Characteristic of most discussions about biotechnology, the topics covered by the BSC were of a very technical, scientific nature. In order to effectively communicate the technical details across the various regulatory jurisdictions, the BSC formed the Biotechnology Regulations InterService Committee (BRIC) (Cantley, 1995).

By the late 1980s, three horizontal directives were being developed in order to establish integrated, Community-wide regulatory measures clarifying the use of biotechnology in the EU. The first two legislative efforts were focused on establishing measures for the contained use of genetically modified microorganisms to ensure protection of human health and the environment. These legislative efforts resulted in Council Directive 90/219/EEC of 23 April on the contained use of GM microorganisms (European Council, 1990a) and Council Directive 90/679/EEC on the protection of workers from the risks of exposure to biological agents (revised by Council Directive 93/88/EEC). Unlike the first two, the third legislative effort targeted the environmental release of GM microorganisms resulting in Council Directive 90/220/EEC of 23 April 1990 on the deliberate release into the environment of genetically modified organisms (GMOs) (European Council, 1990b). The focus of the following discussion will be on Directives 90/219 and 90/220.

With respect to Directive 90/219, both DG III (Industry) and DG XI (Environment) were responsible for developing the legislation; they were *co-chef de file*. There are two reasons for the importance of *chef de file*. First, the relevant DG(s) is responsible for and has considerable scope in setting the policy parameters, such as the legal basis for the legislation. Secondly, the *chef de file* then presents the draft directive to the corresponding Committee at the Council of Ministers. For instance, DG XI would present legislation on behalf of the entire Commission to a subgroup of the Council of Ministers, the Council of Environment Ministers. The advantage is that the DG and the corresponding Council Committee often share ideological and philosophical perspectives.

There were significant philosophical differences between the directorates regarding the regulation of biotechnology. Reconciling these differences required compromise (Patterson, 2000). For example, DG III and DG VI were both initially successful in exempting specific product categories traditionally under their jurisdiction from the draft directive – that is, 90/219 would only apply to those products not covered under existing product legislation, meaning that 90/219 would be a default process-based regulation relevant only when a product-based regulation was not in place. Yet these compromises were vulnerable, because DG III had much fewer resources to devote towards the legislative development than did DG XI (Cantley, 1995). There are two consequences of this. First, DG XI changed the legal basis of the directive from Article 100a to Article 130r, meaning that the goal of the directive was not to facilitate the internal market, but to protect the environment. Secondly, DG XI dropped the compromises struck with DG III and DG VI excluding traditional product categories. Further, these unilateral amendments prevailed, because the draft directive was then put before the Council of Minister's Committee of Environmental Ministers. Therefore, despite the fact that the directive is supposed to represent the entire Commission, in reality directives can be dominated by particular directorates and their philosophies, which are then supported by the corresponding committee at the Council (Patterson, 2000). Indeed, it has been argued that even the BRIC had little influence upon the final directive, because neither it nor the BSC were decision-making bodies; they were only consultative groups, which could be, and sometimes were, ignored (Cantley, 1995).

The final Directive 90/219 was focused on protecting Community health and the environment from the contained use of GM microorganisms. Similarly to the 1976 NIH guidelines, 90/219 divided research risks into two groups. Also, new facilities must notify competent national authorities of their intention to conduct genetic-modification research, so that the authorities can ensure that appropriate protection measures are in place and that DG XI is notified. The directive also permitted member states to consult as widely as they deemed appropriate to incorporate the social dimension in research approvals. Finally, the directive contained provisions about mandatory responses to emergency situations.[2]

The Directive 90/220 on the deliberate release into the environment of GMOs was intended to provide a comprehensive, horizontal, process-based, regulatory oversight, ensuring uniform internal-market conditions while promoting human and environmental health in all member states. The scope of the directive included GM plants, animals and microorganisms, where environmental release for either field-testing or commercial applications required an environmental-risk assessment. It shared many developmental aspects with 90/219, except for one major difference: DG XI was sole *chef de file* for this legislation. Similarly to 90/219, the legal basis of 90/220 was also changed from Article 100a to Article 130r (Patterson, 2000). Also, despite some early compromises with DGs III and VI over products such as medical products, veterinary products, animal feeds and foodstuffs that required risk assessments under product-based legislation, there were no product exemptions from the final legislation. The compromises were dropped and the relevant clause was changed in the Directive to read:

> Consent may only be given for the placing on the market of products containing or consisting of GMOs provided that ... the products comply with the relevant Community product legislation and the products comply with this part of the directive, concerning the environmental risk assessment.
>
> (European Commission, 1988b)

Therefore, GM products faced regulations from both vertical product legislation and from the horizontal regulations under 90/220 before

[2]For a legislative summary see: www.europa.eu.int/scadplus/leg/en/lvb/121157.htm

they could be placed on the market (Patterson, 2000). Similarly to Directive 90/219, Directive 90/220 was also presented to the Council's Committee of Environment Ministers. The Directive outlines that:

> Any person wishing to undertake a deliberate release of a genetically modified organism (GMO) must submit a notification to the competent authority of the Member State within whose territory the release is to take place. This notification shall include a technical dossier of information including a full risk assessment, appropriate safety and emergency response measures and, in the case of products, precise instructions and conditions for use.[3]

Further, according to both the Directives, once a product is approved under either 90/219 or 90/220, a GM product cannot be restricted, prohibited or impeded by individual member states for deliberate environmental release or market placement.

The original Directive 90/220 had two important exemptions. First, it did not include products derived from GMOs, only live GMOs. It was intended that vertical or sectoral legislation would be developed to deal with products derived from GMOs. Secondly, the directive did not initially mention labelling requirements for the consumer's right to know. The regulatory requirements for labelling and packaging were amendments to the directive.

It is important to note that neither 90/219 nor 90/220 considered any risk category as being so great as to require a moratorium. It has been argued that this meant that, despite the strong environmental influence, the directives did not accept speculative risk as a basis for regulations and hence did not go as far as many environmental non-governmental organizations (NGOs) would have liked (Wheale and McNally, 1990). In this sense, although the original directives supported precaution in the application of biotechnologies, they did not support an outright ban. With respect to precaution, Article 16 of 90/220 allows member states to unilaterally impose a provisional ban given:

> justifiable reasons to consider that a product which has been properly notified and has received written consent under this directive constitutes a risk to human health or the environment, it may provisionally restrict or prohibit the use and/or sale of that product in its territory.

An early amendment to the proposed Directive 90/220, which proposed a 5-year partial moratorium on environmental release, was defeated by a margin of only one vote in the EP in June 1989 (Dickman, 1989; Schmid, 1989). Also, the report by the EPs Committee on the Environment called for strict legal liability applying to firms conducting deliberate releases. This would mean that any

[3]Legislative summary: www.europa.eu.int/scadplus/leg/en/lvb/121158.htm

individual or organization claiming for damages due to environmental release would not have to prove that the responsible agent acted negligently. Instead, the claimant simply has to show that the damage was caused by the actions of the releasing agent. The logic is that those releasing GMOs, knowing that they are subject to strict liability laws, will be more diligent in ensuring that the release is safe. This type of liability is similar to US liability and was supported by a 1989 report of the UK Royal Commission on Environmental Pollution, which went further by recommending also that released GMOs should contain a unique and publicly registered genetic marker, so that the source of organisms in the environment can be readily identified (Royal Commission on Environmental Pollution, 1989).

According to the Directive 90/220, the entry point for seeking EU approval is through a competent national authority of a member state, known as the rapporteur, chosen by the company making the submission. The competent authority must review the application, including the technical dossier with the full risk assessment, within 30 days and provide a direct recommendation to the Commission for approval or rejection, along with an application summary. The Commission then circulates the rapporteur's recommendation, along with the application summary, to all member states for a 60-day comment period. Member states may hold public consultations to discuss the application. In the event of approval by all member states, the rapporteur grants deliberate-release or product-placement consent to the applicant. In the event of rapporteur approval but a rejection by another member state, the Commission takes the final decision, according to an Article 21 Committee of qualified majority voting. Consent is binding on all member states.

The two Council Directives, 90/219 and 90/220, were met with significant criticism from scientists, from industry and from trading partners, especially the USA, but also from social interests. Scientific communities claimed that, after nearly two decades of research, there was no evidence that recombinant DNA (rDNA) techniques were associated with an increased level of risk and hence there was no justification for technology-based regulations. Moreover, they claimed that both directives were based on non-scientific criteria and were unduly burdensome to the research process (Cantley, 1995). Essentially, the scientific community argued that scientific rationality should prevail over social rationality.

Industry participants were also very critical of the horizontal, technology-based directives. In June 1989, several pioneering biotechnology firms in Europe formed the Senior Advisory Group on Biotechnology (SAGB), in order to lobby at the European level for legislation conducive to biotechnology research and development (Patterson, 2000). Although the SAGB was unsuccessful in altering the directives, it was notable for its horizontal lobby efforts. This group was especially concerned with the scope of regulatory consultation beyond a science basis.

Other countries, especially the USA, were concerned about the significant departure of the EU regulatory approach from the regulatory principles supported by the USA, along with Canada and the OECD. Both directives represented a deviation from the initially similar regulatory trajectory developed in North America and within several international organizations. The worry was that divergent approaches would yield divergent regulatory outcomes, whereby products approved for deliberate release and commercial use in one jurisdiction would not be approved in another jurisdiction. Indeed, this worry has become reality for GM crops approved as safe in North America.

In response to the negative reaction in 1990, the Commission established another interservice body to consider the criticisms and to reform and coordinate the EU biotechnology policies. The Biotechnology Coordinating Committee (BCC) identified that the directives lacked focus on commercial issues relating to the functioning of the internal market. Essentially, the competitiveness issues associated with biotechnology were identified. This was outlined, for example, in two papers. The first was a communication from the Commission to both the Parliament and the Council entitled *Promoting a Competitive Environment for Industrial Activities Based on Biotechnology Within the Community* (European Commission, 1991), largely drafted by the BCC. There were three main goals:

1. To ensure a coherent regulatory approach and an efficient and simplified interaction between sectoral and horizontal legislation (to eliminate the dual regulations);
2. To ensure that existing legislation is kept under review to reflect rapid developments and technical and scientific progress;
3. To streamline testing and authorization procedures required for biotechnological products.

The second paper was the 1994 White Paper on Growth, Competitiveness, and Employment (European Commission, 1994a), which included biotechnology as a key sector. This paper called for:

1. Constant review of the dynamic developments;
2. Regulatory harmonization with established international practice;
3. Greater public investment.

Yet critics of biotechnology also argued that the directives and these two papers placed too much emphasis upon competitiveness and too little emphasis upon the 'social dimensions' of biotechnology. For instance, a crucial criticism of both directives was that neither included provisions for the public dissemination of information or provisions for public inclusion in the decision-making process (Haerlin, 1990). Therefore, despite the explicit broadening of the remit of the directives to include the social dimensions of biotechnology and the provisions for member state consultation on approvals, there was still significant dissatisfaction with the degree of public input in the approval process. Further, it has also been argued that governments within the EC are too focused on international competitiveness and are devising their science and technology policies to meet this goal. As a result, the public sector is claimed to be too involved with the private sector in promoting biotechnology and the traditional public-protection roles are forgotten (Haerlin, 1990).

From these two fronts the pressures to change the regulatory approach outlined in the two directives was great, and yet the pressures to facilitate international competitiveness were stronger. Hence, given the negative reaction from scientists, industry and trading partners and given the apparent adverse impacts upon European competitiveness, reform of Directive 90/219 became a priority of the Commission and the remit of the BCC. Interservice consultations were held as well as consultations with industry, environmental groups, scientific organizations and trade unions. The first amendment to 90/219 was drafted by the BCC and became Council Directive 94/51/EC (European Commission, 1994b).

This amendment was very contentious. The Commissioner for the Environment, Ritt Bjergaard, displeased with the focus on competitiveness and the internal market, drafted amendments, which were much more sympathetic to consumer concerns about safety and environmental protection, regardless of the risk basis for these concerns. In fact, this proposal very much reflected a social-rationality paradigm to regulating biotechnology. This proposed amendment was presented to the College of Commissioners. This led to a counter-proposal by DG III and DG XII, which was passed by vote in December 1995.

The second amendment to 90/219 was Council Directive 98/81/EC of 26 October 1998. The objective of this amendment was to reconcile 90/219 with the scientific knowledge of and expertise in the risks that had been acquired since the directive first came into force. There were three main amendments.[4] First, the administrative procedures for applying for and receiving approval for contained research were simplified. Secondly, the requirements for notification were linked to the graduated scale of risk, essentially making low-risk research much easier. Thirdly, based on the significant accumulated knowledge about GMOs, a list of GM microorganisms that pose no risk to human health or the environment was included in the legislation.

[4]For a legislative summary: www.europa.eu.int/scadplus/leg/en/lvb/121157.html

In June 1994, the Commission announced its intention to review 90/220; however, this intention did not enjoy the same urgency as that of reviewing 90/219, for two reasons. First, DG XI was sole *chef de file* of this directive and did not share a passion for 'competitiveness-based' biotechnology regulations. Secondly, at this time, there was little European experience with the deliberate environmental release of GM crops, since their development was not as advanced as in North America; the timeliness of the policy issue seemed to be further off.

The first amendment to 90/220 was Commission Directive 94/15/EC of 15 April 1994 (European Commission, 1994b). Similarly to the amendments of directive 90/219, the first amendment was motivated by desires to increase the focus of the directive on European competitiveness in biotechnology and on enhancing the internal market.

An important issue for the market-placement aspects of 90/220 is the labelling issue, as the labelling of products derived from GMOs was, and remains, an important social issue in Europe. The initial 90/220 directive did not deal with labelling. When the first approved GM varieties of soybean from North America began to reach the EU, the position of the EU was that mandatory 'labelling is only necessary when the final product has been substantially modified, which is not the case with "Round-up-Ready" soya' (*World Food Chemistry*, 1996). In other words, the initial position was a hazard or safety basis for substantially equivalent GM crops and not a consumers' right-to-know basis. There was considerable dissatisfaction among consumer organizations, retailers and industry regarding the lack of concern for the consumers' right to know whether GMOs had been used in the product. These groups demanded a mandatory labelling policy for the consumers' right to know about the use of biotechnology in final goods, regardless of the safety or substantial equivalence of the GM product.

The regulatory reaction to these demands was the Novel Foods Regulation (EC) No. 258/97 (European Parliament, 1997). This regulation took 4 years to develop and was the first community-wide labelling legislation designed to meet the consumer's right to know. The intent of the Novel Foods Regulation was to broaden EU regulatory oversight to include products derived from novel GM material, not just products containing viable novel GM material. This was a product-based regulation, where the assessment of safety remained the jurisdiction of 90/220 and the Novel Foods Regulation was only for mandatory labelling requirements. There was considerable support for the Novel Foods Regulation in both the European Council and the Parliament (*World Food Chemistry*, 1996).

As the name indicates, the Novel Foods Regulation was intended to deal with novel foods and food ingredients 'whose nutritional value, metabolism or level of undesirable substances had been significantly changed by the production process' (European Parliament, 1997). Such products were required to be labelled, where the label must include: reference to how the product had been changed to make it novel; the presence of ingredients that might affect health (e.g. allergens); the presence of ingredients that might give rise to ethical concerns; and the presence of a GMO to meet the consumers' right to know. The attempt was to develop scientifically based mandatory labelling requirements consistent with international obligations and designed to convey neutral and transparent information about novel products. Specifically, in respect of the presence of novel GM material, mandatory labelling was required where the presence of modified protein was certain or where its absence could not be certified (as in the case of non-segregation). The Novel Foods Regulation encouraged the voluntary labelling of products that certainly do not contain novel GM material, and, similarly to 90/220 Article 16, the regulations allow member states to impose a unilateral ban 'if new information or a reassessment of existing information suggests a risk or hazard to either human health or the environment'.

However, there are two important aspects of the original Novel Foods Regulation. First, the regulation is not applicable to food additives, flavourings or extraction solvents, which may all be GM material, because other community legislation dealt with these issues. Secondly, as discussed in Chapter 2, GM crops can be novel plants or alternatively they can be non-novel and substantially equivalent to non-GM crops. As a result, the Novel Foods Regulation did not explicitly address the consumer's right to know about the use of GM crops in foodstuffs, only the use of novel GM crops.

The second amendment to 90/220 represents DG XI's reaction to the Novel Foods Regulation, especially to the gap whereby non-novel or substantially equivalent GM crops would not be labelled based on the consumers' right to know. This amendment was Commission Directive 97/35/EC of 19 June 1997 (European Parliament, 1997), enacted under DG XI's emergency allowance to amend Community legislation. Annex III of the directive sets labelling and information-notification requirements for all products made from GM material, whether novel or not. This Annex was essentially designed to override the vertical product legislation on novel foods developed by DG III, with the horizontal regulations of 90/220. DG XI aimed to fill the gaps by requiring mandatory technology- or process-based labelling of all GM products approved for market placement in the EU, not just novel products, by calling for mandatory labelling of GM products approved after 18 June 1997. DG XI wanted specific labelling rules for foods produced from GM crops, not just general labelling rules for all novel foods, because novel foods are not necessarily foods produced from GM crops. Further, under Directive 97/35/EC, firms were encouraged to voluntarily adopt the labelling requirements laid out in this directive for products made from the GM soybean or maize crops approved prior to 18 June 1997. Essentially, Annex III of 90/220 shifted the mandatory labelling policy from a novel, product-based focus to a process- or technology-based focus.

In September 1997, Council Regulation 1813/97 was adopted, effective 1 November 1997, which required mandatory labelling for the GM soybean and maize. Thus DG XI used this amendment to dominate European product legislation with the horizontal biotechnology regulations.

The third amendment to Directive 90/220 was IP/97/1044 of 26 November 1997. This was another labelling amendment, which was focused on establishing the precise requirements referred to in the emergency Directive 97/35/EC. It included specific and mandatory label-content rules.

Together, the Directives 90/219 and 90/220 and the Novel Foods Regulation are the three principal legislative texts dealing with GM crops in the EU. However, there are other important regulations to identify, such as the general EU labelling laws, as well as vertical, product legislation relevant to GM crops.

COUNCIL REGULATION 1139/98: THE SOYBEAN/MAIZE REGULATION. The broad context of European foodstuff-labelling law is outlined by the Council Directive 79/112/EEC. This directive is intended to harmonize the legislation on the mandatory labelling of foodstuffs so as to ensure that products can move freely in the internal market and that consumers are informed and protected. The directive outlines that the label must contain, among other information, a list of the product ingredients – a requirement that is relevant for GM foods. In fact, follow-up work on the directive has been driven by continued concern over the use of GM soybean and maize not considered novel and approved prior to the 18 June 1997 date established by the amendment to 90/220. Council Regulation (EC) No. 1139/98 was established to replace the *ad hoc* measure Council Regulation 1813/97. It addresses the mandatory labelling of certain foodstuffs produced from GMOs, where 'certain foodstuffs' are those containing DNA resulting from genetic-modification techniques, instead of the protein requirement outlined in the Novel Foods Regulation. It supported earlier efforts to develop mandatory labelling requirements for the consumers' right to know. The regulation claims that mandatory labelling requirements are:

> necessary to ensure that the final consumer is informed of any characteristic of food property, such as composition, nutritional value or nutritional effects or the intended use of the food, which renders a food or food ingredient no longer equivalent to an existing food or food ingredient.

In other words, this labelling legislation was increasingly shifting to the social-rationality paradigm, although it maintained the substantial-equivalence principle.

Regulation 1139/98 requires harmonized rules for the additional and specific labelling of foods and food ingredients produced from GM soybean or maize that still contain GM DNA or rDNA. Under the regulation, the following indications were initially required to appear on the label:[5]

1. 'produced from genetically modified soya (maize)' where the product consists of more than one ingredient;
2. 'contains ingredients produced from genetically modified soya (maize)' where an ingredient is designated by the name of a category.

However, the statement that a product 'may contain GM material' was later determined to be unacceptable.

Similarly to both the directive 90/220 and the Novel Foods Regulation, the soybean/maize regulation outlines possible exemptions from harmonization. For instance, it permits member states to unilaterally derogate from the requirements in order to protect human health, the environment or industrial or commercial property.

At the 21 October 1999 meeting of the European Council's Standing Committee for Foods, two notable amendments to the soybean/maize regulation were approved (*EU Food Law Monthly*, 1999; *World Food Law Monthly*, 1999b). First, in respect of labelling, the official EU policy on mandatory labelling supports the consumer right to know about the presence of GM material in food products. Key to such a mandatory labelling regulation is the tolerance level for 'adventitious contamination', which is the acceptable presence of a maximum amount of detectable GM material, where over that amount a product would have to be labelled as containing GM material. The Committee approved a 1% tolerance threshold with a commitment to review the tolerance level as new detection technology is developed and comes into use. These amendments were approved with a vote of 14/15 in favour (with one abstention, Spain). The EP was dissatisfied with this Committee's approvals for two reasons. It wanted to see the regulation establish a specific risk-assessment procedure attached to labelling approval and it pursued a 0.1% threshold level (European Parliament, 1999) – a level also supported by environmental NGOs such as Friends of the Earth. Inversely, the Confederation of Food and Drink Industry wanted a 2% threshold level. In comparison, Japan recently approved a mandatory labelling law with a tolerance threshold of 5% of adventitious contamination.

The second amendment to the soybean/maize regulation was the addition of genetic-modification-derived additives and flavourings

[5]From legislative summary: www.europa.eu.int/scadplus/leg/en/lvb/121090.html

included under the mandatory labelling scheme (*EU Food Law Monthly*, 1999). This proposal was approved with a vote of 12/15 (with two against, France and Spain and one abstention, Ireland). The key to this amendment is that genetic-modification-derived additives and flavourings must be labelled if they:

1. Are, contain or consist of GMOs;
2. Give rise to particular safety (allergy) or ethical concerns;
3. Are not equivalent to their conventionally produced counterparts (i.e. when containing protein or DNA resulting from genetic modification);
4. Are used in 'sufficient quantity', where sufficient quantity is likely to mean anything over the 0.1% threshold (European Parliament, 1999).

In this sense, the regulation is entirely focused on the process and production methods of the final food products.

COUNCIL DIRECTIVE 94/114/EC (AMENDED 70/524/EC). DG VI has also developed product legislation relevant to agricultural biotechnology. First, there is the development of a Novel Feeds Directive, under the Council Directive 94/114/EC, outlining the measures governing the proper use of animal feeds. The motivation for this legislation is derived from the fact that over 80% of the soybean imported into the EU goes into animal feedstuffs and, given the significant proportion of soybean that is GM, it is clear that animal feedstuffs include GM soybean along with non-GM soybean. The amending regulation 70/524/EC introduced new categories of feed additives, including feed containing or consisting of GMOs. This legislation is unique, because it is a process- or technology-based regulation implemented in a vertical regulatory agency.

Secondly, DG VI is developing a Novel Seeds Directive, which is proposed product legislation pertaining to the release of GM agricultural crops. Decision 94/730/EEC established simplified procedures for the release of GM agricultural crops, while Commission Directive 98/95/EC of 14 December 1998 established terms and conditions for a Community-wide seed registry for GM varieties of crops in official categories.

A. EU regulatory administration

Administratively, DG XI is the responsible institution for the horizontal regulations 90/219 and 90/220, including all amendments. In reality, DG XI is the dominant actor in the EU regulatory approach and hence it is reasonable to assume that Community consumer and environmental protection will remain crucial regulatory objectives, despite the competitiveness pressures from DG III, VI and XII.

Along with DG XI, the regulatory role of DG XXIV has also increased. The Consumer Policy and Health Protection Directorate administers the relevant consumer health and scientific committees which advise in regulatory development. The relevant expert advisory committees for GM crop issues include the following:

1. Scientific Steering Committee, which oversees all of the expert scientific advisory committees;
2. Scientific Committee for Foods;
3. Scientific Committee for Plants;
4. Scientific Committee for Animal Nutrition;
5. Scientific Committee for Pesticides.

For instance, the Scientific Committee on Plants is responsible for the technical issues relating to plants for human or animal consumption and the production or processing of non-food products as regards characteristics liable to affect human or animal health or the environment, including pesticides. The Directorate's Consumer Action Plan 1999–2001 has four objectives: first, the development of a community-wide scientific basis for regulatory development under recognized risks; secondly, when risks are unknown, regulatory development should employ the precautionary principle; thirdly, regulatory development must involve consultation with the Institute for Health and Consumer Protection; and, fourthly, food-safety policy should include broader social and ethical considerations. The problem with the first two objectives is their ambiguity. They do not specify the type of unknown risk, hypothetical or speculative. If the focus is on unknown but hypothetical risks, then the subsequent interpretation of the precautionary principle would be the scientific one. On the other hand, if the focus is on unknown but speculative risks, then the interpretation of the precautionary principle would be the social one. The remaining two objectives address 'social dimensions'.

Both DG III and DG VI remain responsible for their relevant product legislation; however, as was shown with the amendment to 90/220 following the Novel Foods Regulation, DG XI's horizontal directives must be considered as dominant to the vertical product legislation.

The EP must be considered as an increasingly important actor in EU regulatory development as well. For instance, in 1998, the Parliament voted 407 to 2 to censure the Commission's approval of a European variety of GM maize and to suspend imports of all GM maize. In late 1999, the Parliament Environment Committee Chairperson, Caroline Jackson, released a resolution (14 December, Strasbourg) calling for joint EP and Council of Ministers decision-making power on issues pertaining to GMO regulations. The basis is that the Parliament is the only directly elected body, which gives it authority over issues of public concern. However, like other EU developments, this resolution further politicizes the issue of food safety in the EU.

All the directorates are subject to the advice and guidance of various committees (*Nature Biotechnology*, 1996). For instance, Community rules on consumer-protection and food-safety issues are subject to the advice of the Scientific Committee for Food. Established under Commission Directive 97/579/EEC, this committee is charged with providing excellent, independent and transparent advice to the Commission. The committee must be consulted on consumer-health and food-safety issues by the Commission. However, the committee may also, on its own initiative, provide advice for the Commission on regulatory issues considered to be within its remit.

Along with the Scientific Committee for Food is the Advisory Committee for Foodstuffs, established under Commission Decision 80/1073/EEC. The Advisory Committee is charged with ensuring close cooperation between the Commission and all relevant Community stakeholders in food-law issues. Similarly to the Scientific Committee, the Advisory Committee is either consulted by the Commission or it provides the Commission with advice on issues relating to the harmonization of all foodstuffs legislation. The fact that the Advisory Committee was established two decades ago indicates the commitment of European food legislation to the whole range of consumer interests and concerns. An important difference from the Scientific Committee is that the Commission is not legally required to consult the Advisory Committee, because the advice is not considered to be independent, as it is with the Scientific Committee.

Member states' scientific bodies may also provide advice and guidance for the Commission on issues of food legislation through the framework established by Council Directive 93/5/EEC. The competent national authority must send to the Commission a list of institutes in the member states that may provide technical assistance in areas including biology, microbiology, biotechnology and novel foods and processes, methods of analysis and risk-assessment techniques.

B. Current regulatory developments in the EU

On 23 February 1998, the Commission presented a proposal for an EP and Council Directive again amending Directive 90/220 (European Commission, 1998). The aim of the proposal was twofold: first, to make the procedure for granting consent to the placing on the market of GMOs more efficient, certain and transparent for industry; and, secondly, to make the procedure more accountable and open, including greater public access to the decision-making process, in order to build public confidence.

There were several industry-orientated proposed revisions. First, the proposal provided for the establishment of harmonized Community-wide methodologies for risk assessment for environmental release – currently under the discretion of the rapporteur country. Key to this harmonized risk assessment was the specification of two categories of releases. Category I releases were those, in compliance with Annex V, for which there is sufficient knowledge about the safety for human and environmental health and where the release is similar to previously approved releases that have not shown evidence of risk. Notification would be made only to the competent authority of the releasing member state and an annual register of all Community-wide notifications for Category I releases would be published. Under the proposal, the competent authority must respond to the Category I notification within 30 days; however, it is unclear whether or not the notifier must await formal consent before release. Category I releases resemble an approval fast-track procedure, similar to the US Department of Agriculture (USDA)'s non-regulated status. Category II releases were for GMOs for which there is not sufficient prior risk information or regulatory experience. Secondly, the proposals also include clarifying the type and quantity of technical information required for an application, such as personnel training and supervision, monitoring schemes and emergency-response plans. Thirdly, timetables for either approving or rejecting applications were set out, in order to prevent the endless regulatory hold-ups that currently burden the approval process of GM crops. Fourthly, the duplication of regulations under both horizontal and vertical regulations was to be removed, so that a risk assessment performed for one is acceptable for the other.

There were also several proposed consumer confidence-orientated revisions of 90/220. The first was to include a fixed-time market-approval period, after which approval would have to be renewed. The initial time period under consideration was 7 years. Secondly, the compulsory postmarket monitoring of all approved products was proposed. Thirdly, a mechanism to allow approvals to be modified, suspended or terminated where new risk information becomes available was proposed. Fourthly, it was proposed that approvals by the Commission could be overturned by the Council of Ministers by a simple majority vote on the recommendation of the relevant scientific committee, such as the Scientific Committee for Food. In fact, the pro posals included mandatory consultations with the Scientific Committee for Food prior to any Commission approval. Fifthly, the proposed amendments included a provision for full legal-liability legislation for damage caused to human or environmental health. This was proposed because several member states had developed or were unilaterally developing their own liability laws for the environmental release of GMOs, in order to create a credible incentive for private firms to minimize the release risks. Sixthly, there was a proposed amendment to make consultations with the EU Ethics Committee and the broader public mandatory prior to any approvals.

The proposed amendments to 90/220 attempted to strike a very delicate balance. On the one hand, they sought to create a streamlined, stable and certain regulatory process with a scientific basis on hypothetical risk and the removal of regulatory duplication. On the other hand, despite the attempts to establish a rules-based approach, the approval process became more politicized, with Parliamentary voting for approvals and Council voting for the withdrawal of approvals. But can regulations be everything to everyone?

On 11 February 1999, the EP approved the proposals with 68 amendments. The amended proposal was presented to the Council for the establishment of a common position on 26 March 1999 (European Commission, 1999). These proposed changes are indicative of the politicization of the biotechnology issue and the desire to regulate agricultural biotechnology according to its social dimensions.

More recently, EU environment ministers claimed that they had agreed on several aspects of the proposed amendments. First, they were all in favour of eliminating the fast-track procedures proposed under the dual-category notification requirements. Secondly, they wanted a negative list of GM applications that could never be approved in the EU, including GM crops that use antibiotic-resistant marker genes. Thirdly, they agreed to a moratorium on new commercial authorizations of GM crops while amendments to 90/220 were being made. Some members, such as Denmark, France, Greece and Italy, argued that the moratorium should remain at least until amendments can be transposed into member-state legislation, a process that was expected to take until 2002 (*Financial Times*, 1999j). Other members were hesitant to endorse this position because of concerns over its legality both at the European Court of Justice (ECJ) and at the World Trade Organization (WTO) (*Independent*, 1999k). Recall that the position of the environment ministers is crucial, because DG XI is the *chef de file* of the Directive and will take the amendment to the Council Committee of Environment Ministers for approval.

In January 2000, the EU Council of Ministers released its 'Common Position' on the various amendments to the directive 90/220, although France, Ireland and Italy abstained from the Common Position (*AgraFood Biotech*, 2000). The Council accepted all proposals from the Parliament, including some that were not accepted by the Commission. Of particular note was the acceptance of the Parliament's interpretation of the precautionary principle (Article 17), which is essentially a legal definition of the social interpretation. The Common Position also indicated that an EU export and trade position would not be defined until after the January 2000 negotiations of the

Cartagena Protocol on Biosafety. This was an important development, because the EU trade position was explicitly linked with a multilateral environmental agreement, rather than the trade-orientated framework supported in North America.

From a commercial perspective, however, regardless of the official status of the moratorium, there has in essence been a *de facto* ban on approvals in the EU since late 1998 anyway, because there is currently not enough support in the European Council for a majority approval.

An important EU regulatory development in respect of the market access of GM crops occurred in late 1999. The EU Environment Commissioner (DG XI), Margot Wallstrom, signalled that she would ask EU member governments to break the *de facto* EU moratorium on GM-crop approvals and authorize 18 GM-crop varieties held up in the EU regulatory system. The condition for the approvals was that the applicant life-sciences firms must voluntarily agree to adhere to the proposed amendments to Directives 90/220, 258/97 and 1139/98, even though they had not yet been enacted. In fact, the life-sciences firms readily accepted the conditions (*EU Food Law Monthly*, 1999). If member governments fail to agree to lift their unilateral bans, then the EU may find itself compelled to take legal action against the bans, which is not a desirable situation.

With respect to mandatory technology-based labelling rules for products derived from GM material, Community sectoral legislation for GM seeds, animal feeds and pesticides are being developed in order to extend the mandatory labelling requirements into a more comprehensive framework, based on the consumers' right to know. An alternative to the mandatory labelling of products that contain GM material is the Community-wide development of a voluntary labelling scheme for organically produced products. Indeed, despite the controversy associated with the mutual exclusivity of organic and GM crops, the Commission is pursuing 'organic' as an alternative to foods that may have been produced from GM crops. The Commission is currently developing a certification framework, in light of the consumer concerns with GM crops, to ensure that the organic label is used properly and consistently. The EU approach is based on Council Regulation 2092/91, which defines when farm produce may be labelled as organic and when produce may carry an EU organic-quality mark or label. This regulation was amended in the November 1992 Council Regulation 3713/92 requiring an inspection certificate to accompany organic food imported into the EU. This raises important trade issues when the definition of 'organic' means something different in another regulatory jurisdiction.

There are also two proposed EU institutional arrangements that could each play an important role in the regulatory system governing GM crops. The first is the European Foods Standards Agency (EFSA) and the second is the European Veterinary and Phytosanitary

Inspection and Control Agency (EVPICA). The EFSA remains a vague proposal and the scope and depth of its regulatory authority currently remain unresolved. On the one hand, some argue that it must be a transparent European institution with real regulatory power and transnational decision-making ability (*EU Food Law Monthly*, 1999). Others argue instead that it should be a forum for reconciling differences in the risk-analysis framework between member states, but not a substitute for independent risk assessments in the member states. The implication of this is that politics will remain inextricably linked to the development of food-safety and environmental-protection legislation in the EU. In fact, in respect of the proposed EFSA, the EU Food-Safety Commissioner, David Byrne, recently noted that responsibility for political decisions, such as the detail and application of the precautionary principle, will remain with politicians (*Eurofood*, 1999). Of course, according to the social interpretation of the precautionary principle, precaution is a political decision, in contrast with the scientific interpretation of the precautionary principle employed in North America, which attempts to disentangle precaution for safety from precaution for political reasons.

The second proposed institutional arrangement which would have significant influence over GM-crop regulations is the EVPICA (Horton, 1997). The objective of the EVPICA would be to enforce EU regulations in the plant- and animal-health fields. The new agency, already approved by the Council, would be located in Ireland and would take over the present inspection activities of DG VI in other member states and in countries who export to the EU. Like the European Medicines Evaluation Agency in London, the new EVPICA would be independent of other European institutions, such as the Commission or the Council.

In the autumn of 1999, the Commission proposed a Draft White Paper on Food Safety, indicating the political priority that food safety has become on the European regulatory agenda. Indeed, it can be expected that the White Paper on Food Safety will represent a strong European food-safety message, in order to gain public confidence in the regulatory system. For instance, among the notable public-orientated proposals of the White Paper are:

1. The creation of a European Food Safety Agency;
2. Changes to the system of scientific advice from the various advisory committees;
3. Increased and mandatory traceability in the entire European food-chain;
4. The creation of a rapid-alert food-hazard warning system.

With respect to the creation of an EFSA, as previously discussed, the scope and the depth of the agency remain undecided and a bit controversial. For instance, it has been noted that the consumer directorate (DG XXIV) is not keen on an independent agency with real authority, as it would overlap the consumer directorate's jurisdiction (*EU Food Law Monthly*, 1999). With respect to the changes to the standing advisory committees, these committees are currently composed of officials from member states, who vote on key food issues pertinent to their particular remits. The proposed changes would see the creation of a single standing committee for all food issues or, at most, two standing committees (food and animal nutrition and health).

There are several areas where further EU regulatory development is likely to occur. First, it may be expected that the soybean/maize regulation will be extended to apply in full to all GM crops commercialized in the EU. Secondly, given the unilateral member-state action, it is likely there will be pressure for the development of a centralized regulatory review process, although the extent that risk assessment and risk management are both centralized at the EU level remains unknown. Thirdly, an important aspect of a mandatory labelling policy is the clarification of acceptable, standardized and accurate test methodologies for adventitious contamination. Also, standardized enforcement and monitoring protocols must be established.

7.3 Agricultural Biotechnology in the UK

In this section, there is an examination of the regulatory approach of the UK because it has experienced the most recent and dramatic rise of consumer concern about GM crops and subsequent pressure to adapt the regulatory approach to social issues.

The regulatory approach to biotechnology in general and agricultural biotechnologies in particular in the UK had initially been very proactive and appeared to be effective at balancing technological progress with precaution. Despite the apparent effectiveness, the regulatory approach has been heavily criticized by social interests, such as consumer and environmental organizations, generating public doubt about its ability to protect consumers and the environment. In fact, it will be argued that social interests have pulled the UK regulatory approach off its initial scientific-rationality trajectory and towards a social-rationality trajectory. The result has been the creation of significant regulatory instability in the UK.

There are two sections to this discussion. First, there is an examination of the evolution of the UK regulatory approach and a discussion of its current structure. Secondly, there is a simultaneous examination of the dramatic rise in consumer concerns about GM crops in the UK and the regulatory changes and proposals that have come about because of this rise in concern.

7.3.1 UK regulatory approach

The first UK advisory group for biotechnology regulations was the Genetic Manipulation Advisory Group (GMAG), established in 1975. The GMAG was under the jurisdiction of the Department of Education and Science and was responsible for monitoring research involving genetic-modification techniques. The GMAG had a focus on self-regulation (Patterson, 2000). The primary contribution of the GMAG was to establish a 'graduated' risk-assessment scheme for various types of rDNA research (Ager, 1990). The GMAG employed the scientific interpretation of the precautionary principle in an attempt to balance the need for control and the need for continued research. For instance, hypothetical risk assessments were precautionary, in the sense that the parameters and assumptions about risk were deliberately risk-averse, to ensure that they overestimated risk. The GMAG relied upon a network of local biological-safety committees to assess the risks of and to monitor rDNA experiments conducted within their jurisdictions.

The first regulatory oversight was under the Health and Safety at Work Act (HSW) 1974, which had general provisions that employers take those actions as reasonably practicable to protect employees and the environment. The Health and Safety Commission (HSC) of the Health and Safety Executive (HSE) are the responsible agencies for ensuring that the provisions of the HSW 1974 are adhered to. In 1978, the Health and Safety (Genetic Modification) Regulations were developed under the authority of the HSW 1974. The regulations called for the mandatory notification of all contained genetic-modification work to the HSC. These regulations continued the initial scientific-rationality trajectory of GMAG.

In 1984, the GMAG was disbanded and replaced by the Advisory Committee on Genetic Manipulation (ACGM) under the authority of the HSE. The ACGM, although not under the direct jurisdiction of any government department, provided advice and information on technical and scientific safety issues associated with biotechnology. The ACGM developed guidelines for those researchers employing genetic-manipulation techniques through working groups chaired by an ACGM member and composed of 'specialists'.

In 1989, the HSC issued new genetic-manipulation regulations in order to deal with the imminent environmental release of GM material. The relevant regulation is under Part VI of the Environmental Protection Act 1990. The basis for the Act was the Royal Commission on Environmental Pollution, chaired by Lord Lewis (Royal Commission on Environmental Pollution, 1989). The Act set provisions for the establishment of the Advisory Committee on Releases to the Environment (ACRE), which is a statutory and independent committee of experts appointed by the Secretary of State. ACRE provides advice for the HSE and the relevant government departments and agencies on the environmental- and human-health and safety issues relating to the deliberate environmental release of GMOs.[6]

In 1991, the UK-based Council for Science and Society released its expert-panel report on the benefits and risks of genetically engineered organisms (Fincham and Ravetz, 1991). The Council was established in 1973 to provide independent and authoritative expert advice for policy-makers and the public on the implications of scientific innovation on society. The report concluded that UK regulations had effectively managed scientific progress with social precaution: 'the effectiveness of a proper attitude is confirmed by recent experience of planned releases. Where public concern has been respected, as in this country, there have been no problems; elsewhere the record is mixed' (Fincham and Ravetz, 1991). Of course, by the late 1990s this conclusion was proved to be premature.

The Deliberate Release Regulations under the Environmental Protection Act were amended in 1992 (SI 1992/3280) and 1995 (SI 1995/304) in order to transpose EU Directive 90/220 into UK legislation. These modifications required environmental risk assessment, notification and consent from the Secretary of State for Agriculture prior to the release or marketing of GMOs. An advisory committee of scientific and medical experts – the Advisory Committee for Novel Foods and Processes (ACNFP) – advises the Secretary of State on applications for consent.

In November 1994, the UK government held public consultations on plant biotechnology, administered by an independent steering committee established by the Science Museum, in order to gauge the social dimensions of biotechnology. The participants chose the questions to ask and the experts to pose them to. The final report was published through the British Biotechnology and Biological Sciences Research Council. The public technology assessment dealt with a wide range of concerns, from human and environmental health and safety to socio-economic impacts and ethical and moral implications (Madden, 1995).

Currently, the UK regulatory approach involves both the Department of the Environment, Transport and the Regions (DETR) and the Ministry of Agriculture, Fisheries and Food (MAFF). The approach is coordinated by the Cabinet Subcommittee on GM Technology. In order to release GM crops for field testing, a Deliberate Release Form must be secured from DETR, which receives advice from the ACRE . The cost of each application is £3000. An application can be renewed or varied for a £700 fee; however, a variation cannot include a change in the size of the field trial or in the duration of the

[6]See www.environment.detr.gov.uk/acre

field trial. A recent field-trial consent for the planting of a herbicide-tolerant rape-seed was revoked because it was found to have contravened the conditions of the variation process. Agrevo, the applicant firm, applied for and received approval for the variation of a previous field-testing consent. But Agrevo should have filed for a new field-trial consent because the intended variations were to increase both the size and the duration of the trial (*The Times*, 18 September 1999). Also providing environmental advice is English Nature, a government advisory body on general wildlife issues.

For commercial approval to use a GM crop in the food supply, the firm must apply to MAFF, while the application is considered by the ACNFP, which is the UK competent authority. This is the UK transposition of the Novel Foods Regulation 258/97. If the GMO is to be marketed alive or viable (e.g. a GM tomato), then additional approval must also be secured from the DETR, which is the UK transposition of the 90/220 directive. In January 2000, the ACNFP and the Advisory Committee on Animal Feedingstuffs (ACAF) held a joint meeting to examine the overlap of their respective remits in respect of assessing novel foods and novel feeds, particularly GM foods and feeds. They agreed to develop complementary assessment approaches.

Finally, the labelling of novel products and products derived from GM material is under the jurisdiction of the Food Advisory Committee (FAC). In 1983, the FAC, chaired by Sir Colin Campbell, concluded that there was no justification to insist on labelling from a safety point of view, and it did not support labelling for the consumer's right to know. Initially, the ACNFP was responsible for establishing UK labelling requirements for GM products, according to the EU Novel Foods Directive. However, early in 1999, this responsibility was passed to the FAC. Anti-GM critics claimed that this proved the lack of expertise of the ACNFP and thus signalled the incompetence of the UK regulatory system (*Daily Telegraph*, 1999). On the contrary, however, this was a reasonable decision, because the remit of the ACNFP is to ensure the safety of novel foods, not to establish labelling policies predicated on the consumers' right to know. Indeed, the former activity is a risk-assessment function, while the latter is a risk-management function, and the two should have been separated. In 1998, the UK developed a voluntary labelling scheme for producers to label foods containing GM DNA. In 1999, the government announced the schedule for mandatory labelling of food produced with GM material. Food retailers, such as supermarkets, had to label the presence or possible presence of GM material as of 19 March, while restaurateurs, café owners and landlords (e.g. pubs) had to do the same by 19 September 1999.

In 1997, the Centre for the Study of Environmental Change at Lancaster University argued that the UK regulatory approach 'appears to be failing to reflect important public concerns arising from acknowledged uncertainties and unknown potential risks from the technology (however low their probabilities)' (Grove-White *et al.*, 1997). To remedy this situation, it was suggested that the UK must actively consult the public through, for example, statutory obligations and that the risk-analysis framework must be replaced by broad consultations based on ascertaining the social intelligence, not the scientific intelligence, of risk.

The UK was the first country to approve a genetically engineered organism for use in the food supply (Harlander, 1993). The approval, in March 1990, was for a strain of baker's yeast (*Saccharomyces cerevisiae*) that produces elevated levels of carbon dioxide. Further, this GMO was determined to be substantially equivalent; hence, there were no labelling requirements. More specifically, in respect of approved GM crops, three varieties of GM crops have been approved, including one GM variety of herbicide-tolerant rape-seed, one GM variety of herbicide-tolerant soybean and one GM variety of tomato, which have been approved for commercial release in the UK by the DETR on the basis of advice provided by the ACNFP.[7] The GM soybean, approved in 1994, was Monsanto's herbicide-tolerant Round-up-Ready variety. The basis for these approvals was very much the scientific-rationality approach.

Yet, following the scientific-rationality approach, the ACNFP has also rejected a commercial application for a GM maize. The Commission approval for Ciba-Geigy's GM maize was rejected, even though it had been approved by the rapporteur country, France. The ACNFP rejected it over concerns regarding the use of the antibiotic-resistant marker gene, because it was felt that this was a safety issue – at least a secondary safety issue – in the sense that the use of antibiotic-resistant genes in crops could add to the growing antibiotic resistance of pathogens harmful to human health. The ACNFP concluded that the GM maize was safe for use in food products where the GM DNA and protein would be processed out; however, in respect of use as a raw animal feed, there was recognized uncertainty about the safe use (ACNFP, 1995). Despite this concern in the UK, the EU Scientific Committees for Food and for Animal Nutrition both assessed the risk as remote or virtually zero and approved the GM *Bacillus thuringiensis* (Bt) maize for use in the food supply and as an animal feed. Labelling was deemed insufficient, because antibiotic resistance affected all and was not a consumer-choice or right-to-know issue.[8]

[7]Regulatory Developments in Biotechnology in the United Kingdom: www.oecd.org/ehs/ukreg.htm

[8]See: Scientific Committee for Food (1996) Opinion of the Potential for Adverse Health Effects from the Consumption of Genetically Modified Maize, expressed on 13 December 1996; and Scientific Committee for Animal Nutrition (SCAN) (1996) Report of the SCAN on the Safety for Animals of Certain Genetically Modified Maize Lines Notified by Ciba-Geigy in Accordance with Directive 90/220 for Feedingstuff Use, expressed on 13 December 1996.

There have been two important developments in the UK regulatory approach. The first development was the formation of an industry association designed to deal proactively with consumer concerns about GM crops. The Supply Chain Initiative on Modified Agricultural Crops (SCIMAC) was informally developed in 1997 and formally established in June 1998. The three main objectives of the SCIMAC are to:

1. Establish industry-wide protocols (codes of practice) for best practice in the commercial introduction of GM crops;
2. Provide practical guidance for growers on the management of specific agronomic traits, such as herbicide tolerance;
3. Maintain an active and open dialogue on GM-crop issues with other interested parties.

SCIMAC is composed of farmers, seed traders, agrochemical firms, plant breeders and biotechnology firms and its creation was driven by the exogenous rapid adoption of GM crops in North America and the endogenous lack of public confidence in the UK regulatory system. The intent was to find an equitable balance in regulatory approach between an industry-led approach and the overly precautionary approach of most GM critics. The approach was based on a successful 30-year programme of preserving the purity and quality of certified seed. In other words, the need to preserve the identify of crops is not a new phenomenon, and sophisticated and successful strategies to achieve this have been in place for decades, although the public was generally not aware of this. To date, several SCIMAC initiatives are worth noting. First, it has announced a voluntary moratorium on the commercial release of GM crops until such time as the government regulators have fully assessed the field trials for safety. Secondly, SCIMAC has developed a code of practice on the introduction of GM crops for both field-testing and future commercial release. Thirdly, it has proposed a farm-scale ecological monitoring programme comparing GM and non-GM crops in the intensive agricultural system, in order to address the concerns of the most vocal anti-GM critics. This programme would be jointly administered by the DETR, MAFF, English Nature and the Royal Society of Plant Breeders.

The second important development in the UK regulatory approach was the establishment of the Food Standards Agency (FSA). The FSA emerged out of the bovine spongiform encephalopathy (BSE) crisis as an attempt to institutionally separate the functions of consumer protection and industry promotion. A public body of 12 independent members from public-interest backgrounds governs the FSA in an open, transparent and consultative fashion. The regulation of GM crops falls under the jurisdiction of the FSA, rather than under the jurisdiction of DETR and MAFF. The FSA implements and assesses the safety of novel foods and processes and has approval authority over releases into the environment.

7.3.2 UK opposition to GM crops and regulatory responses

As a result of recent public resistance to GM foods, the UK regulatory approach has been reformed and amended in an attempt to stem the public concern. The opposition in the UK is extremely polarized, with social-interest organizations, such as Friends of the Earth and the Soil Association, arguing that there is no middle ground: the government must choose either GM crops or organic production. The key point is that the recent public resistance has put pressure on the UK regulatory approach to shift towards a social-rationality approach.

Much of the public concern appears to have arisen from claims made by Dr Arpad Pustzai in August 1998 about the damage that GM potatoes had caused to the health of rats. Dr Pustzai, a researcher at the Aberdeen Rowett Institute, was researching the effect of GM potatoes, engineered with a lectin protein from the snowdrop plant, on the health of rats. This research effort was driven by an apparent lack of scientific evidence on the health effects of GM crops. The early research results, reported by Dr Pustzai, were that the GM potatoes damaged the immune systems, stunted the development of internal organs and led to defective brain development in the rats. After reporting his preliminary conclusions, Dr Pustzai faced harsh criticism from the government, from the Rowett Institute and from other scientists. The scientific criticisms were from those who believed that he broke the rules of scientific disclosure, while the government felt that his preliminary comments undermined public confidence in the regulatory approach. Dr Pustzai's conclusions became the justification for a broad range of concerns about the consumer and environmental impacts of all GM applications, rather than just concern about the safety of GM-lectin potatoes.

Concerns about GM crops have produced a range of reactions, from cautious acceptance to outright rejection, across a broad range of groups. Environmental organizations include Friends of the Earth, Greenpeace, Earth First!, This Land is Ours, English Nature, Gardeners GMO Group and the Royal Society for the Protection of Birds. Consumer groups include the Consumers' Association and Consumers' International. Other organizations include the Soil Association, Christian Aid, the National Federation of Women's Institutes and the British Medical Association. Also, notable opposition came from both the Prince of Wales, a long-time supporter of organic farming practices, and the *Independent* newspaper, with its 'Stop GM Foods' campaign launched on 7 February 1999.

An interesting aspect of the opposition to GM crops or 'Frankenstein food', in the UK is the portrayal of GM crops as 'Monsanto's science'.[9] Yet some of the most advanced GM-crop developers are European firms, such as Agrevo (Germany), Astra-Zeneca (UK), Hoechst (Germany), Novartis (Switzerland) and Rhone-Poulenc (France). In fact, the most controversial EU approval was for Ciba-Geigy's (Novartis) GM maize and the 'illegal' field trials in the UK were undertaken by Agrevo.

In February 1999, the UK government announced the mandatory technology-based labelling of the presence or possible presence of GM material in a food product for food retailers and small food-service establishments. Food retailers, such as supermarkets, had to label the presence or possible presence of GM material as of 19 March, while restaurateurs, café owners and landlords (e.g. pubs) had to do the same by 19 September 1999. While this was the UK's transposition of the EU labelling requirements, it was also importantly driven by the severity of the domestic opposition to GM crops.

Opposition to GM crops also came in the form of commercial threats to those producers who might seek to plant them. For instance, the Soil Association warned that producers would lose their organic certification. In fact, a field trial of Agrevo's GM rape-seed was destroyed by the farmer for this very reason (*Metro*, 8 June 1999). Also, the Institute of Chartered Surveyors advised farmers that growing GM crops could threaten the value of their land and put producers at risk of legal-liability action in the event that environmental damage could be proved.

At the end of March, environmental protesters destroyed field trials of GM rape-seed in Devon, England. Such direct action had occurred previously, in July 1998 in Oxfordshire, but was taken by a more radical environmental organization, GenetiX Snowball. The difference now was that the mainstream environmental lobby, not just the radical fringe, was willing to take direct action and destroy the field tests. For instance, Lord Melchett, the Chairman of Greenpeace, was arrested for participating in the direct action. Lord Melchett defended the illegal action on the grounds that, basically, more than a decade of lawful protest had not imposed their beliefs on the regulatory approach and they were dissatisfied with this fact.[10] Despite the illegal destruction of GM crops and vandalism of agricultural lands, there has been little willingness by the authorities in the UK to seek prosecution against the 'eco-warriors'.

[9]See especially *Independent* (12 October 1999) 'GM Foods: The Debate, A Special Report Produced in Association with Iceland'; and also *Mail on Sunday* (1999); *Express* (18 September 1999) 'Sound Taste in Food'; *Independent* (1999).

[10]The frightening extension of a 'Greenpeace Argument for Illegal Action' is its adoption by other groups, such as right-wing fascist groups, that illegal action is acceptable if 10 years of protesting does not impose their beliefs on the regulatory approach.

In May 1999, the Commons Environmental Audit Select Committee, an all-party committee examining environmental policies, recommended four immediate policy responses to the public concern. A moratorium on commercial plantings of GM crops, tighter rules on field trials, broader regulatory consultations encompassing ethical issues and more public funding of GM-crop development to counteract the perception that it was a private technology (*Independent*, 1999c). The basis for the recommendations was that the increasing public concern would create a loss of public confidence in the regulations. Therefore, in order to be socially responsive, this recommendation called for tighter regulations to assuage public concerns, regardless of whether or not the concerns were driven by speculative risks.

In defence of the scientific UK regulatory approach, Professors Janet Bainbridge, Chairperson of the ACNFP, and John Beringer, Chairperson of the ACRE, testified to the Commons Science and Technology Committee that a ban on the field testing or commercial cultivation and sale of GM crops would not increase consumer and environmental safety. Instead, its only impact would be a severe impact on the competitiveness of the UK GM-technology industry (*Independent*, 1999a).

In May, a government review of the UK regulatory approach was released. The review was authored by England's Chief Medical Officer, Professor Liam Donaldson, and by Sir Robert May, the government's Chief Scientific Adviser. The review concluded that 'there is no current evidence to suggest that genetically modified technologies used to produce foods are inherently harmful' (*Financial Times*, 1999f). However, it also called for the creation of two new advisory committees and a National Surveillance Unit to rebuild confidence in the regulatory approach. The two new advisory committees are the Human Genetics Commission and the Agricultural and Environment Biotechnology Commission. The key feature of the two new advisory committees is the deliberately broad participation base. The National Surveillance Unit is proposed to monitor the long-term health and nutritional impacts of GM material in the food supply. Critics of the review included the Consumers' Association, which raised concerns about the review's failure to deal with long-term risks, the British Medical Association and the Friends of the Earth, which called the report 'miserably inadequate' (*Daily Telegraph*, 22 May 1999).

Also in May, the Royal Society, the UK's senior scientific academy, announced that the independent review by six scientists of Dr Pustzai's research found that the conclusions claimed were not supported by the evidence. The Royal Society report concluded that the research was 'flawed in design, execution and analysis' and that there was 'an incorrect use of statistical tests'. It also concluded that no conclusions could be drawn about the safety of GM foods in general from research on one animal species, one type of food, one type of gene and one genetic-transfer technique. Further, the report noted that the difficulty with scientific innovation is that 'nothing can be absolutely certain in a field of rapid scientific technological development'. Yet this uncertainty has given rise to public fear, which has 'been echoed, and sometimes cynically distorted, by tabloid newspapers and special interest lobby groups' (*Daily Telegraph*, 22 May 1999; *Independent*, 1999d).

In September 1999, Sir Richard Sykes (1999 President of the British Science Association) called for secret trials of GM crops in order to advance the technology, while protecting it from direct, illegal action or vandalism. He argued that, without the trials, Britain would fall behind in capacity and competitiveness.

In September 1999, Friends of the Earth successfully challenged the DETR over the field-trial consent given to Agrevo for its GM rapeseed. Indeed, Agrevo had violated the conditions for when an applicant can make a variation to a field-trial consent and when an applicant must apply for a new field-trial consent. When the illegal action was discovered, Agrevo had already planted three of four fields, and the intended fourth planting was put on hold. While admitting that Agrevo had violated the law, the Environment Minister (Michael Meacher) argued that the previous three trials should continue because it was a technical error and not a health and safety issue. The reaction by the Friends of the Earth was that the UK government was willing to bend the law to support biotechnology companies (*Express*, 18 September 1999; *The Times*, 18 September 1999). Yet the UK government has 'bent the rules' to the benefit of critics as well, as it has not pursued legal action against protesters who destroyed legal field trials of GM crops.

Also in September, Friends of the Earth announced that research by the National Pollen Research Unit, in Oxfordshire, UK, revealed that pollen from GM crops can be found up to 3 miles (4.8 km) away from the field test site. At issue is not the fact that cross-pollination can occur; instead, it is what kind of tolerance is demanded. In Chapter 5, section 5.1.6, it was noted that there exists controversy between those who support zero tolerance and those who support some minimum tolerance of risk. Friends of the Earth, along with the Soil Association, support a zero-tolerance policy and regard the research findings as completely unacceptable. In defending the regulatory policies, Professor Allan Gray of the ACRE acknowledged that there is a minimum risk of cross-pollination and that the field-testing regulations were designed to minimize that risk to an acceptable level, not to eliminate it completely. Further, he points out that safety is the real issue and, since the GM crop is safe, there is no safety concern if it cross-pollinates.

In October 1999, the British medical journal, *The Lancet*, announced that it would publish Dr Pustzai's complete study on the effects of GM-potato consumption on the immune system of rats. This was heralded as a vindication of Dr Pustzai's claims (*Independent*, 1999r). Yet, far from a vindication of Dr Pustzai, this represents a triumph of good science and an example of why the scientific methodology must be used in dealing with risk assessment. It has been noted that 'good science is difficult to achieve, and bad science is all too easily hyped into a scare story' (*Independent*, 19 May 1999). The original evidence of harm was based on incomplete and impartial research findings and was subsequently denounced by other scientists. In this case, Dr Pustzai has provided *The Lancet* with the complete research, which has been peer-reviewed and found to be credible. Yet, as noted by the Royal Society, the fact remains that this is research about one genetic-transformation technique on a potato and from a specific plant (snowdrop) not currently used on any commercialized GM crops. It is not evidence that GM techniques in general are unsafe.

In October, the UK government announced an initiative to embark on broad participation negotiations on the future regulatory approach to GM crops, including the participation of consumer and environmental groups, the biotechnology firms and government (*Independent*, 3 October 1999). In fact, regulatory development in the UK can be expected to be focused on assuaging public concerns about GM crops through the risk-management procedure of risk analysis, rather than through the hypothetical risk-assessment process. For instance, the regulatory negotiations are to be facilitated by the Environment Council with a view to finding common ground between the supporters and the critics – a political balance.

On 13 October 1999, the government announced that the ACNFP would routinely disclose non-commercially confidential information on novel foods under assessment on a website prior to making its recommendations. Further, once an assessment is complete, there will be a public-comment period before it is forwarded to the European Commission. This development is aimed at enhancing the transparency of the assessment process and ensuring that the UK recommendation on novel foods reflects public concerns more fully.

On 5 November 1999, the UK Environment Minister Michael Meacher announced that, instead of a ban on commercial plantings, a voluntary agreement with the GM-crop developers and seed companies to resist commercial-scale plantings until 2002, at the earliest, had been reached. The industry, represented by the SCIMAC, developed a 'Code of Conduct' for producers using GM crops. The Code of Conduct includes advice on the distances at which GM crops must be grown from either non-GM or organic crops. Further, these voluntary guidelines are underpinned by Technology Use Agreements between the GM-seed companies and the farmers. This agreement allows for 4 full years (1998–2002) of field-trial evidence to be collected and assessed. However, GM crops grown elsewhere and approved as safe in the UK will be allowed to be imported and used in the food supply.

Therefore, the UK regulatory approach, initially on a scientific-rationality trajectory, can be expected to focus primarily on the speculative risks fuelling public resistance to GM crops, as it shifts to a social-rationality approach. The implication of this shift is that, as long as GM crops remain a highly politicized issue in the UK, the regulatory approach will remain unstable and unpredictable. The paradox is that attempts to answer the very questions about GM crops that the critics are basing their resistance on are being thwarted by vandalism in the name of 'direct action'. Indeed, it appears that the critics are willing to undermine due scientific process because of the threat that it will contravene their interests in the exact same way that they accuse multinational biotechnology firms of doing.

7.4 EU–Member-state Regulatory Integration

The EU–member-state strategy for regulatory integration, according to the 'new approach', is a strategy of deeper social integration blending regulatory coordination with regulatory competition. The basis of this strategy is the establishment of essential minimum requirements (EMRs), harmonized across all member states. For controversial issues, the EMRs may be decided by the Council of Ministers or by the ECJ (Woolcock, 1996). An important feature, then, is that, in the absence of agreement among all member states, the EU EMRs are top-down regulatory principles imposed on the member states.

On this foundation, there is a process of *ex ante* regulatory coordination through mutual recognition. This has been an effective strategy for removing the politics from many regulatory integration efforts; however, this strategy has not been successful in all areas, especially GM-crop regulations.

When member-state regulatory action contravenes policy positions of the EU, the European Commission may bring legal action against the member state in the ECJ. The member may be ordered to pay fines for failing to comply with Community rules. Lister argues, in respect of food-safety regulations, that the ECJ has made it clear that:

> by becoming members of the European Community, Member States had given up their sovereign authority to write food laws as they saw fit. Rather, they had only limited ability as Member States to restrict marketing within the EU of foods so long as they meet minimum essential requirements.
>
> (Lister, 1992)

Member-state regulatory divergence has been exemplified by Austria and France, who have both imposed unilateral bans on GM crops, which have been determined to be unjustified at the EU level. For instance, Austria imposed unilateral bans on Ciba-Geigy's (Novartis) GM Bt maize in December 1996 and on Monsanto's GM Bt maize (MON810) in May 1999, on the basis of evidence presented to the European Commission's Scientific Committee for Plants. This evidence allegedly indicated a link between GM maize and environmental damage, and the action was taken under Directive 90/220's Article 16. In October 1999, the Scientific Committee ruled that Austria's ban was not justified under Article 16 and Austria was expected to lift the ban, although no time frame was set for this. Similarly, in an opinion, the ECJ indicated that the French ban on GM Bt maize was unjustified.

In short, unlike the degree of regulatory integration between Canada and the USA, based on shared regulatory principles, regulatory integration between the EU level and member states is not necessarily a foregone conclusion, because the approaches can differ significantly. For instance, the examination of the EU and the UK regulations reveals that it is inaccurate to think of the 'EU' as a homogeneous regulatory jurisdiction. Instead, awareness of the limitations of EU–member-state regulatory integration is crucial.

7.5 Conclusions

The EU regulatory structure is summarized in Table 7.1. The dominant actor is the environment directorate, DG XI, while the consumer directorate, DG XXIV, also plays an important role. This means that social interests, the clients of these directorates, can be expected to exert more influence over the EU regulatory development process than economic interests. Legislatively, the horizontal, process-based directives 90/219 and 90/220 dominate over vertical, product-based directives and regulations.

Given the mix of horizontal and vertical legislation, the breadth of regulatory participation and the uncertainty surrounding the many proposals for revision, the current EU regulatory framework for GM crops is enigmatic and unstable. Essentially, the EU is attempting to redesign the regulatory system to meet the spectrum of concerns from scientists and industry about research and development and competitiveness to the whole range of consumer interests and concerns. Further, it must do this with the highly politicized issues associated with agricultural biotechnology specifically and with food-safety and environmental issues in general.

Similarly to the regulatory approaches in the USA and in Canada, the EU regulatory approach may be categorized according to the

Table 7.1. EU agricultural biotechnology regulations.

Agency	Regulatory oversight
Horizontal legislation	
DG III and DG XI	Council Directive 90/219: On the Contained Use of Genetically Modified Microorganisms Process-based
DG XI	Council Directive 90/220: On the Deliberate Release of Genetically Modified Organisms into the Environment Process-based
DG III	Regulation 258/97: Novel Foods Novelty-based labelling scheme for the consumers' right to know
DG XI	Annex III of 90/220: Mandatory labelling legislation Process-based for the consumers' right to know
Product legislation	
DG VI	Council Directive 94/114 amended 70/524: On Feedingstuffs Process-based
DG VI	Novel Feeds Regulation Novelty-based, substantial equivalence
DG VI	Novel Seeds Regulation: establishing terms for the registration of GMO varieties Novelty-based, substantial equivalence

scientific-and social-rationality paradigms (Table 7.2). The most important point is that despite an initial scientific rationality trajectory, internal factors have caused the EU approach to shift to a distinctive social rationality trajectory flowing from the tradition of regulatory accountability or political control over discretionary regulatory power.

The regulatory belief is one of technological precaution, where it is not possible to disentangle actual from perceived risks. Instead, perceived risks are a legitimate regulatory target, despite the lack of scientific justification for such a perception. For example, in 1997 the EU approved a GM variety of maize that was initially approved by the rapporteur member, France. However, in the meantime, French elections brought in a new political administration less supportive of GM crops. The new Prime Minister overruled the recommended approval of the French competent authority, causing the chairperson of the authority to resign. Four GM-crop applications have been approved as safe by scientific committees but have been rejected by politicians, fearful of the public backlash, should they endorse the GM crops.[11] In

[11]These crops are: Monsanto's GM sugar beet and maize, along with two varieties of GM oil-seed rape developed by Agrevo.

Table 7.2. EU regulatory-approach check-list.

	Scientific rationality	Social rationality
General regulatory issues		
Tradition		✓ Regulatory accountability
Belief		✓ Technological precaution: socially responsive regulations increase public credibility and confidence
Type of risk		✓ Recognized ✓ Hypothetical ✓ Speculative
Substantial equivalence		✓ No: technology-based
Science or other factors in risk assessment		✓ All four regulatory hurdles in risk assessment
Burden of proof	✓ Traditional: minimize Type I (rejected when safe)	
Risk tolerance	✓ Minimum	
Science or other factors in risk management		✓ Social dimensions of biotechnology, including other legitimate factors in risk management
Specific regulatory issues		
Precautionary principle		✓ Social interpretation
Focus		✓ Technology-based
Structure		✓ Horizontal, new structures
Participation	✓ Judicial decision-making	✓ Wide, social dimensions
Mandatory labelling		✓ Mandatory, technology-based for the consumers' right to know

fact, in November 1999, EU member governments voted against scientific approvals for commercial introduction. The vote was taken in an Article 21 committee and the rejection was based on a claim that more information was required (*Public Ledger*, 1999b).

Moreover, it has been argued that GM crops are enmeshed in the political and social world of European food politics, where social interests exhibit relatively more policy power than in North America.

> The process of [food safety] policy formation in Europe is a considerably more complex and nuanced system than appears to be the case in the USA. The power of vested interests and social democratic impulses makes the final policy position taken in Europe a far less scientific and rigid proposition than is expected in the Sanitary and Phyto-sanitary Agreement [which] contains a rather fundamentalist model of regulation

> that is out of kilter with the political economy of regulation in a European environment.
>
> (Evans, 1999)

In general, 'Europe has a history of prohibiting new food technologies that are used widely in America' (*Economist*, 1996).

The EU regulatory approach that emerges from this context is a technology-based approach. As a consequence, the types of risks in the risk-analysis framework include speculative risks about the technology based on perceptions, and substantial equivalence is rejected as a gateway regulatory principle. Given the rejection of the substantial-equivalence principles, all GM crops would be considered novel and would face all four regulatory hurdles for market approval: safety, health, quality and socio-economic benefit, applied in a technologically precautious manner. To date, the EU approach to the risk-analysis framework has not abandoned the traditional burden of proof (innocent until proved guilty), nor has risk tolerance been abandoned, although there has been considerable pressure to implement a zero-tolerance policy, as was implemented in the EU beef-hormones ban. In terms of the regulatory factors considered under risk management, the EU approach certainly entertains a broad range of social dimensions. Here science may inform the regulatory decision but it does not make the regulatory decision. Indeed, it has been argued that regulating GM crops 'is not done in a vacuum, but the concrete environment of societies with specific preferences' (Haniotis, 2000). Such other legitimate factors include vague concepts such as the multifunctionality of agriculture, whereby any GM crop perceived to pose a speculative risk to the multifunctionality of EU agriculture may be rejected.

With respect to the five regulatory principles for GM crops, the EU approach employs the social interpretation of the precautionary principle. According to this interpretation, the principle is a risk-management tool, where a legitimate use could be to ban a technology based on speculative risks – not hypothetical risks – because of domestic risk perceptions. The regulatory focus is technology-based through a regulatory structure of horizontal regulations.

In terms of regulatory decision-making, the EU approach, thanks to its commitment to the social dimensions of biotechnology, involves wide actor participation. To date, however, the regulatory decision-making has not moved to a consensual basis, as it continues to retain a judicial decision-making process. While conceptually wide actor participation is vital, operationally it is a challenge. It was argued in 1991 that the development of EU agricultural policy was difficult because of the involvement of agriculture and budget, foreign and trade ministers (Phillips, 1991). With GM crops, consumer, environmental, technology and industry ministers can also be added to this mix, further complicating policy development. This wide actor participation contradicts the arguments that decision-making in the EU is closed and 'technocratic'. In fact, it has been argued that social dimensions have been a significant part of biotechnology policy in the EU (Sclove, 1996). Landfried (1997) argued that 'the diverse mixture of national, supranational, technocratic and political interests within the Commission, Council, Parliament and committees' means that regulatory decision-making in respect of biotechnology is not the sole responsibility of a narrow group of technocrats.

Finally, under the dominance of DG XI and DG XXIV, EU labelling policies adopt a mandatory, technology-based labelling strategy, based on the consumers' right to know about the presence of GM material, regardless of the scientific risk assessment of the product and the substantial equivalence of the GM crop with non-GM crops.

Given the EU regulatory approach characterized in Table 7.2, what type of regulatory integration would be supported? As previously mentioned, the DG I Trade of the European Commission has explicitly stated that the EU trade-policy approach to regulatory integration supports the regulatory coordination strategy of the Cartagena Protocol on Biosafety.

As the EU regulatory approach has shifted to a social-rationality trajectory in response to social interests, several key aspects have emerged. First, the EU regulatory approach is unstable, leading to significant commercial uncertainty. As social interests tend to hold the balance of power in the regulatory development process, EU regulations are unstable, as they react to perceived public risks in order to be accountable and socially responsible. Indeed, GM-crop applications approved in North America, in compliance with EU regulations and in fact approved by European scientific committees, have been rejected by politicians, leaving the application in a suspended state of regulatory hold-up.

Secondly, the EU approach has created a regulatory floor, not a regulatory ceiling – that is, member states, according to Article 130r Title XVI Environment of the Treaty Establishing the European Community, can and do unilaterally impose national regulations that are more stringent than corresponding Community regulations. For instance, the UK, France and Denmark have called a partial halt on GMO approvals, while France, Austria, Greece and Luxemburg have imposed unilateral bans on certain new crops. Austria has also imposed both a complete ban on the commercial release of GM crops and a complete ban on the patenting of life. This action was driven by a public petition endorsed by 20% of the electorate (Ho, 1998). Yet, if member-state legislation is in contravention of EU law, then why does the Commission not challenge the legality of this action before the ECJ? Several cases have been considered by the ECJ,[12] but essentially

[12]A French ban on Novartis GM Bt maize was ruled as unjustified in an opinion by ECJ Advocate-General Jean Michel, released in November 1999.

the social-rationality approach and the politicization of the GM-crop issue combine to ensure that, although member states are in contravention, the Commission simply lacks the political support to pursue regulatory integration through legal channels. The result is fragmentation of the single market and considerable uncertainty surrounding the approval of GM products in Europe. Therefore, while the European 'new approach' was developed specifically to deal with the problems of mixing politics with regulatory development, it appears that, in respect of GM-crop regulations, the EU regulatory approach has shifted to a social-rationality paradigm, which necessarily mixes the politics of accountability and social rationality with the regulatory development process.

The third important aspect of the EU regulatory approach is that it appears to fail to meet the demands of either economic or social interests. Similarly to the North American regulatory approach, the European regulatory approach may be assessed according to a commercial and a social perspective. From a commercial perspective, the European approach has failed to establish the stable and certain regulations necessary for industrial development. Approval applications have been paralysed within the complex and time-intensive approval process, frustrating GM-crop developers and creating trade tensions. In fact, the European approach has sacrificed commercial objectives to meet social objectives. Yet, despite this effort, the regulators face severe criticism that they remain too commercially orientated. The result is that the European approach, while striving to find the middle, has instead produced a regulatory approach that fails to meet both commercial and social needs.

Given the regulatory difficulties, would EU food laws be better served as regulations rather than directives? Directives are developed and accepted at the EU level but are then transposed into the legal framework of each member state. In order to aid transposition, directives sacrifice detail in favour of general principles. The result is EU regulatory instability, as the legislation of member states can vary considerably. As an alternative, EU food regulations must be accepted 'as is' by member states and are often more detailed and specific, allowing less possible variation. This, of course, makes the development and negotiation of regulations more onerous, but they are explicitly efforts in regulatory coordination and, once developed, they yield both regulatory detail and stability.

Part IV
Analysis

There are two objectives of Part IV. The first is to bring together Parts I, II and III in order to analyse why the traditional trade-diplomacy approach is incapable of addressing the transatlantic regulatory regionalism associated with genetically modified (GM) crops and, hence, stands at a crucial crossroads. Despite the fact that both regions employ the risk-analysis framework, the North American and the European Union (EU) approaches are drastically different. In comparing the two, the former follows a scientific-rationality paradigm for regulatory development, which then cascades into support for a regulatory-competition strategy to regulatory integration. The latter follows a social-rationality paradigm for regulatory development, which then cascades into support for a regulatory-coordination strategy to regulatory integration. The traditional trade-diplomacy approach is congruent with the North American approach but dramatically divergent from the EU approach and, as such, it is unable to deal with the transatlantic regulatory regionalism associated with GM crops.

The second objective is to take the above analysis a step further and to propose and examine an amended trade-diplomacy approach that addresses the transatlantic regulatory regionalism in an economically and socially acceptable manner. Essentially, the second objective is to identify the type of regulatory approach capable of bridging the fundamental regulatory differences in order to facilitate regulatory integration. Finally, from this 'ideal' regulatory framework follow implications for governments, industry, international governmental organizations and non-governmental organizations, which will be examined in the last section.

8 Transatlantic Regulatory Integration

8.1 Trade Diplomacy at a Crossroads

This study has focused generally on the trade-diplomacy problem of regulatory regionalism and specifically on the problem of transatlantic social regulatory barriers facing genetically modified (GM) crops. It has been revealed that trade diplomacy stands at a crucial crossroads. Maintaining the traditional trade-diplomacy approach, which promotes regulatory development and integration according to the scientific-rationality paradigm, will increasingly erode public support for trade diplomacy and marginalize its ability to facilitate international regulatory integration (the scientific- and social-rationality paradigms are summarized in Table 8.1). The reason for this is simple: while top-down trade-liberalization rules focused on border measures tended to have very little impact upon domestic citizens, top-down trade rules disciplining the use of domestic regulations aimed at ensuring food safety and environmental protection or meeting moral, ethical and religious concerns are perceived to have an enormous impact upon domestic citizens.

Given that a top-down trade-liberalization solution is not feasible, strategies for enhancing regulatory integration must be bottom-up. It has been shown that the domestic regulatory approach for GM crops is, in fact, a function of endogenous political-economy factors. These include various interests and events, along with the traditional regulatory role of the state and the competitiveness of the jurisdiction in respect of the particular regulatory issue. While there is a transatlantic consensus that the risk-analysis framework is the most appropriate framework for developing GM-crop regulations, there are divergent views on how to implement the risk-analysis framework properly.

Table 8.1. Comparison of scientific- and social-rationality regulatory approaches.

	Scientific rationality	Social rationality
General or framework regulatory debates		
Belief	Technological progress Enhances growth and development, leading to higher social regulations – a regulatory race to the top	Technological precaution Technology cannot be separated from socio-economic factors – must be socially responsive
Type of risk	Recognized Hypothetical	Recognized Hypothetical Speculative
Substantial equivalence	Yes	No
Science or other factors in risk assessment	First regulatory hurdle based on identified safety or hazard risk	All four regulatory hurdles
Burden of proof	Innocent until proved guilty Minimize Type I error (rejected when safe)	Guilty until proved innocent Minimize Type II error (accepted when unsafe)
Risk tolerance	Minimum	Zero tolerance
Science or other factors in risk management	Safety or hazard-based Risk management is for risk reduction and prevention only	Broader socio-economic concerns Risk management is for social responsiveness
Specific regulatory debates		
Precautionary principle	Scientific interpretation Risk-assessment tool	Social interpretation Risk-management tool
Focus	Product-based, novel applications	Process- or technology-based
Structure	Vertical, existing structures	Horizontal, new structures
Participation	Narrow, technical experts Judicial decision-making	Wide, 'social dimensions' Consensual decision-making
Mandatory labelling	Two scenarios Substantially equivalent: Safety or hazard basis Novel: Consumers' right-to-know basis	All GM crops are novel: Consumers' right-to-know basis

Economic and social interests compete to influence the regulatory trajectory. The former supports scientific rationality and technological progress, while the latter supports social rationality and technological precaution. Consequently, regulatory debates emerge from these broad regulatory goals. Seven framework debates have been identified, along with five specific debates associated with GM-crop regulatory principles. The result is regulatory instability within jurisdictions, as competing interests influence the policy-makers, the regulators and the general public as to which regulatory trajectory is most appropriate.

Moreover, as the regulatory trajectory is a function of the endogenous political-economy factors, which can be quite different, regulatory approaches can be quite different, resulting in regulatory barriers to trade and regulatory regionalism. With respect to regulatory integration, this study has revealed that a regulatory jurisdiction faces three integration parameters: the level of integration; the depth of integration; and the strategy of integration (Fig. 8.1). For the regulatory regionalism associated with GM crops, the dominant level of integration is the transatlantic level. The integration depth must deliberately target deeper social integration beyond the shallow economic integration favoured in the traditional trade model. The integration strategy must involve proactive regulatory coordination and not be limited to reactive regulatory competition only. From Fig. 8.1, the proposed regulatory integration strategy straddles IIC and IID.

In short, traditional trade diplomacy has arrived at a crucial crossroads. The traditional approach to regulatory integration is represented by area IA (Fig. 8.1). Although this narrow approach to multilateral, shallow regulatory competition has worked effectively in dealing with border measures to trade, it does not deal effectively with social regulatory barriers to trade, because it fails to account for the vital political-economy factors that determine social regulations.

Two scenarios emerge. The first is that trade diplomacy retains its traditional focus on regulatory integration according to IA (Fig. 8.1) and hence assumes a marginalized role in international integration, only dealing with border measures to trade. Of course, this scenario does nothing to solve the problem of regulatory regionalism.

The second scenario is that the trade-diplomacy approach is amended to better facilitate either global or regional regulatory

1. Integration Level

I. Global/multilateral	II. Regional/bilateral or plurilateral

3. Integration strategy

2. Integration depth		Regulatory competition	Regulatory coordination
	Shallow 'economic' integration	A	B
	Deeper 'social' integration	C	D

Fig. 8.1. Integration parameters facing a regulatory jurisdiction.

integration in areas C and D, rather than just A (Fig. 8.1). Amending the traditional trade-diplomacy approach to better account for the social dimensions influencing regulations will enhance the prospects that trade diplomacy can effectively deal with regulatory regionalism and promote international regulatory integration. In other words, as trade diplomacy reaches deep into areas of national competence in order to enhance regulatory integration, it must be sensitive to the political-economy factors shaping the development of the regulatory framework and subsequent regulations within a particular jurisdiction. Essentially, the goal of trade diplomacy must not be to defend the traditional economic-integration approach at all costs. Instead, the goal of trade diplomacy must be to encourage international integration in an economically and socially acceptable manner.

The objective now is to propose and examine how transatlantic regulatory regionalism may be overcome. In the remainder of Chapter 8, the features of an 'ideal regulatory framework' will be discussed (section 8.2), followed by a discussion of the implications and relevant policy recommendations (section 8.3).

8.2 Ideal Regulatory Framework

Following from the analysis of Parts I, II and III, it is now argued that a regulatory framework combining the pragmatism of the scientific-rationality approach with the responsiveness of the social-rationality approach could address the integration challenges of social regulatory barriers to GM crops. This so-called ideal regulatory framework must promote regulatory stability and integration, subject to several important parameters. First, the framework must balance the competing economic and social interests, calling simultaneously for technological progress and precaution. In this sense, the approach must achieve an acceptable political compromise.

Secondly, the regulatory framework must be precise enough to be operational – that is, in order to deal with regulatory externalities, eliminate instability and avoid trade tensions, the regulatory framework must have precise rules that clearly outline regulatory procedures. It should avoid vague or ambiguous regulatory principles, which lead to delay and indecisiveness.

Thirdly, the regulatory framework must be strategically targeted. The regulatory resources of governments are crucially limited and yet technological innovation is a dynamic phenomenon. In this sense, the regulatory framework must be a dynamic approach, focused on the application, management and distribution of GM technology and capable of keeping pace with technological innovation, rather than a complete command and control approach, which quickly becomes obsolete.

Fourthly, in respect of regulatory integration, the regulatory framework must avoid autarky. It must disentangle those regulatory aspects that can be coordinated with regulations in other jurisdictions from those regulatory aspects that, because of crucial domestic pressures, must remain under the complete control of the jurisdiction. Further, it must be considerate of relevant international obligations.

Given these important parameters, the ideal regulatory framework employs a risk-analysis framework to the regulation of GM crops. As identified in Part III, both the North American and the European regulatory approaches are within the risk-analysis framework, and yet they differ substantially on several important framework principles and specific principles for regulating advanced technologies such as GM crops.

Comparing the North American and the European regulatory approaches reveals the sources of regulatory regionalism. Essentially, while the North American regulations remain congruent with the scientific-rationality approach, the European regulations have departed from this trajectory and shifted to a distinctly social-rationality approach to regulating GM crops, which tends to be more congruent with its regulatory tradition. Table 8.2 compares the two approaches over the framework and specific regulatory principles identified in Chapter 5.

The comparison in Table 8.2 reveals that the transatlantic regulatory regionalism and social regulatory barriers facing GM crops are caused by structural differences in the regulatory approaches employed in North America and Europe. In order to promote stable regulatory integration, the ideal regulatory framework must overcome the significant structural regulatory differences in the risk-analysis approach employed in the two dominant jurisdictions.

Although both risk-analysis approaches have their strengths, neither approach is the most appropriate for addressing the issue of social regulatory barriers to GM crops. For instance, the North American approach, while supported by economic interests, is simply not socially responsive enough to address social interests and would be vulnerable to a dramatic public crisis over GM foods, as has occurred in the UK. Yet the European Union (EU) approach, which has shifted to meet the demands of social interests, has failed to establish an approach supported by either economic or social interests. It does not have a solid enough foundation of risk assessment to be stable and predictable and is constantly reacting to perceived risk, regardless of the scientific justification for the public concern. Therefore, both approaches must be amended to more equitably address the legitimate economic and social interests in a stable, transparent and socially responsible manner.

The ideal regulatory framework aims to be socially responsive in the regulatory development process, but not to the detriment of regulatory framework stability and integration. The key to this is to

Table 8.2. Comparison of North American and European regulatory approaches.

	North American approach	European approach
Tradition	Regulatory independence	Regulatory accountability
Belief	Separation of economic objectives from social objectives, because intervention increases costs and decreases productivity, competitiveness and economic prosperity	Market is embedded in social factors, which cannot be separated from economic factors, because intervention increases participation, increasing productivity, competitiveness and economic prosperity
Authority	Independent regulators; fourth level of government	Democratically accountable regulators
Effectiveness	Economic interpretation of market efficiency	Political-economy interpretation of success in meeting social dimensions: social responsiveness
Paradigm	Scientific rationality	Social rationality
Type of risk	Recognized or hypothetical	Recognized, hypothetical and speculative
Substantial equivalence	Yes	No
Science or other factors in risk assessment	First regulatory hurdle based on identified safety or hazard risk	All four regulatory hurdles
Burden of proof	Traditional	Traditional; pressures to reverse
Risk tolerance	Minimum	Minimum; pressures for zero tolerance
Science or other factors in risk management	Safety- or hazard-based	Broader socio-economic concerns
Precautionary principle	Scientific interpretation	Social interpretation
Focus	Product-based, novel applications	Process- or technology-based
Structure	Vertical, existing structures	Horizontal, new structures
Participation	Narrow, technical experts Judicial decision-making	Wide, 'social dimensions' Judicial decision-making
Mandatory labelling	Safety or hazard only: voluntary, market-based for the consumers' right to know	Mandatory, process-based for the consumers' right to know
Integration	Regulatory competition	New approach: Regulatory coordination and regulatory competition

employ the scientific-rationality approach to risk assessment, providing a solid foundation for risk, and yet to employ the social-rationality approach to risk management, providing a socially responsive framework to meet social interests. Beck (1992) argues that despite the apparent mutual exclusivity of the scientific- and social-rationality paradigms, they can be integrated because they are mutually reinforcing: 'scientific rationality without social rationality remains empty, but social rationality without scientific rationality remains blind'.

In short, the ideal regulatory framework essentially builds social credence into the scientific-rationality paradigm. This is accomplished by combining a scientific risk-assessment approach with a social risk-management approach and a transparent risk-communication strategy. This ideal risk-analysis framework will be discussed below in its three constituent parts: risk assessment, risk management and risk communication.

8.2.1 Risk assessment

The objective of risk-assessment procedures is to produce neutral and transparent risk information to inform the risk-management process. Risk assessment can enhance regulatory stability and regulatory integration by producing clear and precise procedures, establishing essential minimum requirements (EMRs) for safety applicable in all regulatory jurisdictions.

In order to integrate the scientific- and social-rationality paradigms in the risk-assessment procedures, the only principle of scientific rationality that must be abandoned is the idea that empirical questions are logically prior to normative questions. The reason for this is that a decision must be made as to the type of risk that the risk assessment will focus on. This is a normative question, which is in fact prior to the empirical question. But, once it is answered, empirical questions may be asked and assessed by recognized experts in the field, resulting in legitimate, credible, neutral and transparent risk information to assist in the risk-management procedure.

With respect to the type of risk targeted in risk assessment, the focus of risk assessment must be on asking empirical questions about hypothetical risks, not speculative risks. In other words, risk should be something that can be defined empirically and testable by empirical means, rather than based on unsubstantiated logical possibilities. Empirical questions have the important ability to clarify risk according to accepted analytical methods and causal–consequence mechanisms, while limiting the social and political influence over the risk-assessment procedure.

Is this approach acceptable? Indeed, even within the scientific community, there is considerable disagreement over empirical conclusions. Yet disputes among hypothetical risks can at least be assessed through further study and analysis. In fact, it has been argued that this is the approach used in medical risk assessment and it enjoys widespread confidence from all stakeholders (Balk, 1993). Speculative risks based on logical possibilities cannot be 'refuted' by the scientific process and are the refuge of those uninterested in a genuine risk dialogue. In this sense, speculative risks are not an operational regulatory principle.

Focusing on hypothetical risks because they are subject to accepted analytical methods does not mean that new analytical methods cannot be developed, nor does it imply that 'no evidence' means 'no risk'. It has been argued that the problem with current risk-analysis approaches to GM crops is the lack of a predictive ecology framework (Rogers, 1990). Without such a framework, critics argue that they must bypass risk assessment based on hypothetical risks and rely instead on risk management that assumes speculative risks. Yet this approach adds nothing to risk information and seems to undermine the development of an alternative analytical method, because of the lack of research and development. Indeed, it would be irresponsible to abandon biotechnology, because biotechnology and genetically modified organisms (GMOs) are an inevitable feature on the industrial landscape. Developments in biotechnology in general benefit specific applications; hence, technology will continue, even with a ban in one specific area, such as agricultural biotechnologies. Also, due to the significant economic stakes of agricultural biotechnology, a regulatory ban in one jurisdiction will only lead to a redistribution of GM-crop development, not a halt. It is in the interest of the most advanced nations, with the most advanced scientific capacity and regulatory systems, to regulate the hypothetical risks of GM crops according to the best analytical methods at their disposal in order to set the global standards.

Further, when scientific risk assessments do not yield conclusive results, it is an imperative part of the scientific precautionary approach that no evidence from inconclusive results implies more investigation, and not a conclusion of no risk. Therefore, the development of new analytical methods, such as predictive ecology, must be a regulatory priority. The new methods may prove that past hypothetical risks were underestimated; however, they may equally prove that past hypothetical risks were overestimated, as they were at the initial Asilomar Conference in 1975.

With respect to the risk-assessment debate regarding the principle of substantial equivalence of some GM crops, there appears to be no scientific or moral reason to assert that production-improving and some output-improving GM crops are not substantially equivalent to non-GM varieties for regulatory purposes. Instead, from a regulatory perspective, risk assessment should be focused on the hypothetical risks of the novel features of GM crops, not on the use of genetic modification *per se*. This is similar to long-standing regulatory control over exotic plants and pests regulated according to their novelty. Hence, from a regulatory perspective, it is reasonable to conclude that GM and non-GM crops are substantially equivalent unless the GM crops are novel. This should not be interpreted, however, to mean that consumers do not have a right to know whether a crop is GM or conventional. Of course, if consumers want to know, they must have access to this information. In this sense, substantial equivalence is only a useful principle for establishing the necessary level of regulatory oversight for risk assessment, but it cannot be a dominant principle for establishing risk-management responses. Instead, the consumers' right to know must be an important principle in risk management, which is discussed in greater detail below (section 8.2.2).

Should risk assessment focus on other legitimate factors? In risk assessment, where the goal is to develop empirical questions to gain neutral, transparent and credible information about the hypothetical risks of GM crops, there is no role for the normative 'other legitimate factors', including food quality and socio-economic impacts. That is, once the definition of risk is established, a science basis predicated on hypothetical risks and the best available analytical methods should be the only factors in risk assessment. Risk assessors should only be charged with ensuring that a food is safe, not assuaging consumer fears and proving social benefit.

With respect to the debate over the regulatory principle of burden of proof, a logical and legitimate compromise is to maintain the traditional burden of proof (minimize the risk that a technology is rejected when it is in fact safe) but regulate new technology more stringently. However, this is then a risk-management activity and will be discussed in greater detail below (section 8.2.2).

Participation in the risk-assessment procedures must be reserved for experts only, and the real issue is finding independent and credible scientific experts to perform the hypothetical risk assessment – a process requiring considerable consensus building (MacKenzie, 1993). The experts must be as independent of industry as possible and can include known supporters and critics. This ensures wide participation among the experts and ensures that broad concerns will remain on the agenda. Yet, at the same time, the involvement of scientific experts decreases the variance in normative preferences. Decision-making by the participants must be judicial and rules-based, not consensus-based, in order to ensure timeliness in risk assessments in a field characterized by rapid technological innovation. Finally, the deliberations and decisions of participants must be fully transparent to all interested stakeholders.

The ideal risk-assessment procedures are more congruent with the North American approach than with the European approach and are generally supported by the Organization for Economic Cooperation and Development (OECD), the World Health Organization (WHO) and the Food and Agriculture Organization (FAO). In this sense, the European approach must shift towards a rules-based and science-based risk-assessment procedure for GM crops, adopting the scientific interpretation of the precautionary principle and limiting the risk assessment to short- and long-term human and environmental safety, not to other legitimate factors. It must also adopt a product- or application-based focus. In addition, it must increase the transparency of the assessment procedures and resist the temptation to overturn assessment decisions that are incompatible with political decisions. Making these amendments solves the European problems of instability and inoperability.

Although largely congruent with the North American approach, the ideal risk-assessment procedures do require the North American approach to include greater participation in the procedures. Experts from a broader range of interests and yet familiar with the specified frame of reference and the analytical methods, would participate and wield policy power, not just the GM-crop developers and government regulators.

The ideal scientific risk-assessment procedures benefit both regulatory stability and integration. Stability is increased, because the language of risk, safety and precaution is harmonized, eliminating confusion and ambiguity associated with divergent risk assessments and effectively levelling the playing-field for all stakeholders. Technological precaution is explicitly built into the assessment procedures, through risk-averse assumptions and likelihood functions, allowing credence goods to shift to experience and search goods in a controlled manner. Non-market regulatory objectives, such as human and environmental safety, would be dealt with by specifying the frame of reference at the outset, establishing regulatory floors based on shared EMRs. Participation would be greatly increased, making critics part of the scientific risk-assessment procedure, rather than excluded actors claiming unsubstantiated risks from outside the regulatory development process.

The prospects for regulatory integration are also increased, because of the harmonized risk-assessment procedures. This limits the problem that could arise because there is no supranational institution to decide on EMRs. It also limits the degree of possible normative disagreement, because scientific disagreements can be dealt with subject to accepted rules of scientific debate. Essentially, the ideal risk-assessment procedures establish regulatory principles that are operational in an international integration framework, because the conditions for risk information are precise and stable.

The likelihood that both the North American and the European regulatory approaches can and would adopt the ideal scientific risk-assessment procedures are good, despite the necessary modifications.

The analysis of the economic and social interests has revealed that, in most instances, opposition to GM crops is not actually opposition to the technology *per se*. Instead, it is opposition to the application of the technology. The ideal scientific risk assessment focuses on the risks of these application issues, rather than obfuscating the debates by widely condemning the technology of genetic modification. Further, in both regions, there is a desire to avoid the 'my scientist vs. your scientist' debates which occurred in the World Trade Organization (WTO) beef-hormones dispute. More recently, this debate occurred in the autumn of 1999, when British beef approved as safe in Britain and by the EU Food Safety Committee was not approved as safe by French authorities and France refused to lift its ban on British beef. Both sides repeatedly claimed that they had science on their side. Adopting the ideal scientific risk-assessment procedures would prevent or drastically limit regulatory barriers being justified according to controversial debates over science.

8.2.2 Risk management

While it is necessary to centralize and harmonize the scientific risk-assessment procedures, the objective of risk management must be to build social credence into the scientific regulatory approach. Although the goal of the risk-management process is risk reduction and prevention, it is undeniably an inherently political exercise of balancing the rights and interests of both supporters and critics. In this sense, it is incorrect to characterize risk management as just a process of creating economic instruments to correct market failures, where the only indicators of regulatory effectiveness are economic measures of market efficiency. In other words, once based on a common framework of risk information, risk-management decisions must retain the freedom to be socially responsive.

An important feature of the risk-management stage is that endogenous or internal pressures lock risk management at the national level. Essentially, the economic and social stakes are too high to give up authority over risk management. Therefore, it is impractical to expect an international regulatory approach to emerge in the short or even medium term. Instead, risk management remains the domain of regulatory jurisdictions and cannot be effectively centralized or harmonized at the multilateral level.

Within this context, the ideal risk-management process aims to ensure that regulatory development, while being socially responsive, is simultaneously considerate of both the scientific risk-assessment procedures and the obligations to other countries under various international agreements.

Under the ideal risk-management process, science matters and the risk-management deliberations must be based on a common foundation of risk information, provided by the risk-assessment procedure. With respect to the regulatory debate associated with risk tolerance, risk management must accept that zero tolerance is not a realistic goal. Instead, the focus must be on the relative risk of the GM-crop application, subject to some tolerated level of risk, permitting the technological progress necessary to increase risk information. Without this basis, risk management is arbitrary and vulnerable both to regulatory capture and to criticisms that it is scientifically unjustifiable. One solution is to introduce a 'technology penalty' (van den Daele *et al.*, 1997) at the risk-management stage, designed to build consumer confidence in the regulatory control over new technologies. In fact, it has been argued that greater scrutiny of GM crops has already been established in this way, without having to reverse the burden of proof. Ager (1990) and Fincham and Ravetz (1991) both argue that GMOs have faced greater regulatory scrutiny in the UK, despite the fact that there is no empirical evidence to justify this. In this case, risk-management regulations are primarily for confidence building to address the credence factors and not for enhancing safety. Yet, given the discussion on the UK (Chapter 7), it is clear that the technology penalty must also be accompanied by a more concerted effort by proponents to address the information gap created by the commercialization of their own applications. In essence, the burden of proof remains the traditional principle, but the new technology must meet more stringent conditions on product or application performance and on dealing with credence factors.

Therefore, unlike risk assessment, science cannot be the final arbiter of risk management; it can only inform the regulatory development process. With respect to the risk-management debate over the use of science or other legitimate factors, the ideal risk management must consider the social dimensions of GM crops and involve the participation of all stakeholders from either end of the support–opposition spectrum. In this sense, risk management cannot be independent or normative-free, because it cannot be disentangled from the interests and concerns of both supporters and critics.

Crucially, this requires compromise and concession on the part of both supporters and critics and the realization that the goal of risk management is not to build normative consensus; rather, it is only to strike an acceptable balance. On the one hand, private firms must endeavour to assess GM crops to the highest level of consumer and environmental protection available. They must also respect the social dimensions and credence nature of GM crops by meeting the consumers' right to know, even in the absence of safety issues. Recall from Chapter 2 that the failure to preserve consumer choice among GM and non-GM crops has had disastrous consequences in Europe, despite the absence of safety issues or justifications. In addition, product-liability laws should be included in regulatory approaches, to make sure that the risk to human and environmental safety becomes internalized by the firm as commercial risk, which the firm then minimizes. Indeed, it has been argued that, with product-liability laws, 'regulators may have significantly more freedom . . . in coercing the market toward efficient outcomes' (Holloway, 1999).

On the other hand, critics cannot expect regulatory approaches to fully reflect their belief systems and must be prepared to compromise without illegal action. They must participate in and respect the risk information provided in the risk-assessment procedure, and some level of risk tolerance must be accepted in risk management, as there is no such thing as zero risk. They must be prepared to engage in cooperative, non-antagonistic actions during the risk-management process.

The ideal risk-management process also requires governments to abandon the inward focus common for food-safety and environmental-protection regulation. Instead, governments must give equal consideration to the impact of their risk-management decisions upon other countries as well, according to their obligations under various international agreements. There is no such thing as 'autarky' and, as regulations have an impact on market access, risk-management policies must consider those impacts. Governments must resist domestic pressures to impose regulatory barriers that are unjustified according either to the scientific risk assessment or to acceptable measures agreed to in, for example, trade agreements.

To achieve this, mutual recognition must be built into the domestic risk-management process. The reason for this is because domestic regulations represent a political balance and, once a balance is achieved, it is crucial for political stability that the regulatory framework prevents foreign GM crops from circumventing the domestic requirements, thus creating concerns that destabilize the balance. If governments cannot resist establishing unjustified regulatory barriers, then they must be prepared to acknowledge that the barriers are driven by unjustified factors in violation of international obligations. Hence, although governments can opt out of their obligations, they cannot do so without cost – in this case, paying compensation to affected foreign firms in a manner that they have already agreed to under international agreements. In this sense, national governments are free to choose whatever risk-management regulatory response they wish, but they are not free to use controversial scientific justifications.

Additionally, the intergovernmental discussions on mutual recognition must abandon traditional trade-diplomacy features, such as limited participation and untransparent, closed-door discussions. Instead, discussions of mutual recognition must occur with both an open dialogue and wider participation. The key result of the open, transparent, mutual-recognition strategy is that, while national governments are free to pursue those regulatory responses that are domestically necessary, they must accept the regulatory responses of other governments as equivalent to their own. Of course, this further supports the centralization of scientific risk-assessment procedures, so that a domestic government can ensure that the safety assessments are domestically acceptable.

The ideal risk-management process is more congruent with the European regulatory-accountability approach, where regulatory development responds to the social dimensions, not just to market failures. In order to achieve the ideal risk-management process, the European approach must fully commit to scientific risk-assessment procedures as an important foundation for risk management, and it must be prepared to support a level of tolerable risk, even in the face of public pressure. Further, social risk-management decisions taken in absence of a scientific justification must be considered in the context of international rights and obligations, where contravention, although permissible, is not without cost.

Meanwhile, the North American regulatory approach must accept the legitimacy of other factors in risk management, beyond scientific factors. In other words, it must accept a social risk-management process. This includes provisions to address the credence nature of GM crops by permitting regulations predicated on the consumers' right to know, even in the absence of safety concerns. For instance, this means that non-product-related production and processing method (PPM)-based labelling rules must be developed and accepted and not challenged under narrow international trade law, which fails to adequately account for a social approach to risk management.

Essentially, in dealing with regulatory barriers to trade, the ideal risk-management approach is crucially focused on building social credence into the scientific-rationality approach. It accepts the fact that regulations are political and that political factors differ between countries and regions, making it impossible to establish international risk-management responses. Divergence in regulatory approach will occur, but this divergence will be minimized by basing risk management on centralized scientific risk-assessment procedures. Further, the proactive regulatory-coordination strategy of mutual recognition prevents the tensions and disputes that arise under regulatory competition.

The ideal risk-management approach can increase regulatory stability by ensuring that it is based on a foundation of credible, harmonized scientific risk-assessment procedures. It preserves sovereignty or autonomy over regulatory responses, meeting crucial subsidiarity pressures. It acknowledges the critical role of political and social factors in regulatory development and it supports the legitimacy of these other factors beyond the limited market-failure and scientific factors.

The ideal risk-management approach also enhances regulatory integration through regulatory coordination by promoting cooperative, proactive efforts to develop mutual recognition agreements, avoiding the forced convergence of regulatory competition. Mutual recognition avoids both the difficulties of harmonizing risk management across countries and regions with divergent economic, social and political factors and the potential democratic deficit created when international institutions wield regulatory authority in the absence of clear accountability provisions. It promotes greater commercial certainty through stable market-access rules, creating economies of scale and reducing transaction costs. Although this strategy cannot prevent regulatory barriers, it can discipline their use in a framework more capable of meeting the concerns that make disciplining regulatory barriers under a regulatory competition framework at the WTO unacceptable. Finally, the ideal risk-management process can prevent the perception of a competitive deregulation or regulatory race to the bottom, because it explicitly provides for the inclusion of legitimate, non-market objectives in domestic regulatory responses.

8.2.3 Risk communication

The biggest failure in the risk-analysis framework in both North America and Europe has been a lack of transparency – the failure to provide adequate and accurate risk communication. GM crops exhibit credence factors that are poorly understood by consumers because of a significant information gap. Both supporters and critics have failed to provide accurate information, and instead have manipulated information to meet their own interests. Consider, for instance, supporters and their global-welfare promises and critics and their sensational dread predictions based on emotive, speculative risks. Governments, pulled in opposite directions to meet the interests of both supporters and critics, have understandably chosen to quietly deal with GM crops in an attempt to keep them out of the political agenda. Together, the polarity of views and the untransparent government approach represent a completely unacceptable strategy of risk communication.

The ideal risk-communication strategy is a complete commitment, by all stakeholders, to the two-way flow of transparent information between scientific risk-assessment procedures and the risk-management process. The risk information produced in the scientific risk-assessment procedures must be available to all interested stakeholders, including interested parties in other countries. To make this possible, GM-crop developers must abandon both their global-welfare promises and their overused claims of 'proprietary information'. Consumers have a right to know, and a lack of transparency is easily construed to imply that there is something to hide. Providing the results of in-house human- and environmental-risk assessments does not reveal crucial proprietary knowledge, but it does significantly limit the vulnerability of firms to spurious, sensational criticisms. Further, the results of the scientific risk-assessment procedures, as well as the deliberations, must be completely available to all stakeholders in the risk-management process. In turn, however, the social dimensions of the risk-management process must also flow back into the scientific risk-assessment procedures – in terms of the scientific precautionary principle, for instance.

The goals of the ideal risk-communication strategy are to reduce the information gap caused by the credence nature of GM crops and to increase public confidence in the regulatory approach. To achieve this, the North American approach must accept that the risk communication of GM crops must meet the consumers' right to know about the non-safety aspects. The European approach must increase its transparency to eliminate the unacceptable regulatory hold-up without explanation. Both approaches must better acknowledge the unavoidable regulatory externalities or global aspects of GM crops. By addressing the credence nature of GM crops, the polarity is minimized, resulting in informed consumerism based more appropriately on the actual benefits and risks, and fully in support of consumer choice. Given the enormous economic, human-health and environmental benefits, it is difficult to believe that consumers, truly informed, would reject GM crops. Instead, they would understand that it is the application that matters and that applications are best controlled through a risk-analysis framework. In short, the ideal risk-communication strategy enhances stability and integration by enhancing transparency and public confidence.

8.3 Conclusions and Implications

The ideal regulatory framework essentially builds social credence into the scientific-rationality approach, resulting in a stable framework that is congruent with deeper social integration according to a regulatory-coordination strategy. Social credence is built in by ensuring consumer information, trust and choice. This approach meets the prerequisites of balancing the competing interests within an operational, dynamic, rules-based approach, capable of managing the applications of advanced technologies.

There are significant implications of this ideal regulatory framework for governments, the GM-crop industry, international organizations and non-governmental actors, which will be discussed below in this order.

8.3.1 Implications for governments

Within the transatlantic regulatory regions, governments must reconcile the divergence in their regulatory frameworks. With respect to risk-assessment procedures, the EU approach must become more like the North American approach, while, for the risk-management process, the North American approach must become more like the EU approach, with greater stakeholder participation. For risk communication, both the North American and EU approaches must endeavour to be more transparent and accessible.

In dealing with the problem of regulatory integration, governments must engage in proactive regulatory coordination, not regulatory competition. In this sense, the North American approach must become more like the EU and could learn a great deal from the European experience with integrating economies. Also, recognizing the severe weaknesses in dealing with regulatory integration, governments must refuse to take disputes over social regulatory barriers to the WTO.

The ideal regulatory framework allows countries to pursue their own social regulatory approaches unilaterally and they may even ban GM crops, if they wish, according to the scientific interpretation of the precautionary principle. However, it prevents them from misusing 'science' in order to support domestic political concerns and holds them accountable for obligations that they have already agreed to under international treaties.

8.3.2 Implications for the GM-crop industry

The GM-crop development industry in both North America and Europe must accept that regulations are not instruments used only to correct market failures and that consumers are not just economic agents. Instead, regulations play a legitimate, crucial role in meeting broader social concerns and they must not be viewed only as economic costs. The industry must take a proactive role in addressing the information gap associated with their products, through accurate and transparent risk communication. The consumer does have a right to know whether or not products are derived from GM crops. In meeting this right, industry must endeavour to identity-preserve, regardless of the incremental costs. As segregation is necessary for both output-trait GM varieties and bioengineered products, this simply speeds up the introduction of effective segregation procedures.

Industry must also accept that science cannot be the final arbiter and, instead, there must be scope for the use of the scientific interpretation of the precautionary principle in regulatory approaches. The scientific precautionary principle is a legitimate regulatory objective and it may be used to restrict or ban particular GM-crop varieties.

Finally, industry must accept that the comprehensive range of concerns associated with GM crops requires an international regulatory integration approach that is broader in remit than the traditional trade approach. In this sense, industry must accept that the WTO cannot be the dominant international institution and they must be prepared to avoid using the WTO to adjudicate on the appropriateness of domestic risk-management regulations. Therefore, industry must engage fully in a regulatory coordination strategy, through, for instance, simultaneous approval applications in North America and the EU, in order to highlight the common, shared objectives rather than the differences.

8.3.3 Implications for international institutions

International organizations must facilitate regulatory coordination where possible and, where not possible, must be prepared to stay out of the way. The key lesson is that, when the fundamental regulatory framework is unstable and divergent in North America and the EU there is very little hope that an international regulatory framework will be imposed top-down. The implication of this is that international organizations, such as the OECD, Codex, WHO and FAO, should continue to assist the risk-assessment procedures for the safety of GM crops, and yet they should avoid involving themselves in the risk-management process, for fear of tarnishing their independence and credibility.

Similarly, the WTO should be kept out of social regulatory integration when what is at issue is the development of fundamental regulatory principles. It cannot credibly rule on which regulatory framework ought to be used, because it is fundamentally built on the economic perspective and yet social regulations are most often driven from the social perspective. This is not to say that the WTO is not a valuable institution. Indeed, it should continue to focus on trade liberalization through the administration and enforcement of rules and, in areas of little controversy, it can play a role in brokering international rules. But it should not be charged with establishing the fundamental regulatory framework and adjudicating whether or not social regulations are compatible with trade objectives. Such a role only serves to seriously undermine support and confidence in an otherwise very valuable institution.

Of course, this raises an important question about what to do with the Agreement on Sanitary and Phytosanitary Standards (SPS Agreement) – a trade agreement right in the middle of the controversy between trade objectives and social regulations. Clearly, the SPS

Agreement should not be renounced, but careful guidelines on its applicability must be established. Essentially, it is vital to determine whether or not the regulatory barrier under investigation has created trade tensions because of disagreements over framework principles or specific principles. If, for instance, there is widespread international agreement on framework principles, then the trade dispute over specific regulatory interpretations can be reasonably brought to the WTO's dispute-settlement mechanism. If, on the other hand, the dispute arises because two jurisdictions disagree on a fundamental framework principle, then it is obvious that the WTO cannot credibly deal with this dispute.

8.3.4 Implications for international non-governmental organizations

The international non-governmental organizations (INGOs) that compose the social interests must accept that regulatory development is a political exercise requiring compromise and concession. While the ideal risk-analysis framework amends the regulatory approaches to more effectively deal with their concerns, at the same time, INGOs must be prepared to participate in the entire regulatory development process, even if their belief systems are not met in their entirety. For instance, criticisms of the risk-assessment procedures can only be considered valid if the organizations are involved in the procedures, rather than raising sensational speculative risks from the sidelines. They must be fully engaged in the risk-management process and accept that their interests are not the only ones that regulations must consider. Finally, and perhaps most crucially, they must also commit to an accurate and transparent risk-communication strategy, with the objective of increasing informed consumerism. They wield enormous power in capturing media attention on emotive issues and this power must be wielded responsibly. In fact, their greatest contribution would be as an active participant in the regulatory development process, engaged in finding common ground among the stakeholders.

8.3.5 Looking ahead

There should be no doubts about the difficulty of meeting the challenge of social regulatory barriers to the trade of GM crops. Dealing with these barriers requires drastic structural changes to the traditional trade-diplomacy framework, as well as crucial compromise and concession from all involved stakeholders – both supporters and critics. This is not an issue of extremes. Instead, all stakeholders must endeavour to find the elusive common ground of a more reasoned and rational approach. Only then can the benefits of trade liberalization and socially acceptable regulatory oversight be simultaneously achieved.

Once the transatlantic rules for establishing a regulatory approach capable of balancing technological progress with precaution are developed, the approach must be taken to the international level to begin the process of multilateralization. Admittedly, this next step is also fraught with enormous challenge, because it must reconcile the often competing interests of developed and developing countries – the so-called North–South divide. In this sense, transatlantic regulatory coordination is time-sensitive, as perhaps more pressing or urgent North–South issues will continue to be insufficiently addressed until the pillars of the international economic order reconcile their transatlantic differences.

Finally, only when a multilateral regulatory framework is established can the WTO then play its role as the principal trade liberator, clear of the controversies associated with social regulatory integration.

References

Ackerlof, G. and Dickens, W. (1982) The economic consequences of cognitive dissonance. *American Economic Review* 72, 307–309.

Advisory Committee on Novel Foods and Processes (ACNFP) (1995) *Report on Approval Application for Ciba-Geigy's Maize*. ACNFP, London.

Ager, B. (1990) The regulation of genetic manipulation. In: Wheale, P. and McNally, R. (eds) *The BioRevolution: Cornucopia or Pandora's Box?* Pluto Press, London, pp. 236–252.

AgraFood Biotech (1999a) No. 19, December. AgraEurope, London.

AgraFood Biotech (1999b) No. 20, December. AgraEurope, London.

AgraFood Biotech (1999c) Political Storm over World Trade Biotechnology Initiative. No. 19, 8 December. AgraEurope Ltd, London.

AgraFood Biotech (2000) No. 21, January. AgraEurope, London.

Agrevo UK Ltd (1996) Position Paper to Professor Derek Burke, Chairman of the UK Advisory Committee on Novel Foods and Processes. 20 November.

Almas, R. and Nygaard, B. (1995) *New Biotechnologies: Attitudes, Social Movements, and Regulation. A Comparative Analysis of a European Discourse*. European Congress of Rural Sociology, Prague.

APHIS (Animal and Plant Health Inspection Service), United States Department of Agriculture (1992) *Genetically Engineered Organisms and Products; Notification Procedures for the Introduction of Certain Regulated Articles; and Petition for Non-Regulated Status*. Draft Guidelines 12 June 1992 57:40632 and Final Policy March 1993 58:17044. Federal Register, Washington, DC.

APHIS (Animal and Plant Health Inspection Service), United States Department of Agriculture (1999) March 1999.(www.aphis.usda.gov/oa/new/ab.html)

Appell, D. (2001) The new uncertainty principle. *Scientific American*. (www.sciam.com)

Arrow, K.J. (1963) *Social Choice and Individual Values*, 2nd edn. Yale University Press, New Haven, Connecticut.

Arthur, W.B. (1989) Competing technologies, increasing returns, and lock-in by historical events. *The Economic Journal* 99, 116–131.

ASA (American Soybean Association) (1996) Position Paper to Mr Vigeveno, Unilever. 9 December.

Balk, R.A. (1993) Public values and risk assessment. In: Fessenden MacDonald, J. (ed.) *Agricultural Biotechnology: A Public Conversation About Risk.* NABC Report 5, NABC, Ithaca, New York, pp. 87–95.

Baltimore, D. (1982) Priorities in biotechnology. In: *Priorities in Biotechnology Research for International Development: Proceedings of a Workshop.* National Academy Press, Washington, DC.

Barach, J. (1999) Comments at the FDA Public Consultation of Agricultural Biotechnology in Washington, DC, 30 November (see *AgraFood Biotech,* 1999a).

Barton, K.A. (1998) Biotechnology: catalyst for change in agriculture. In: Hardy, R. and Segelken, J. (eds) *Agricultural Biotechnology: Novel Products and New Partnerships.* NABC Report 8, NABC, Ithaca, New York, pp. 47–57.

Beck, U. (1992) *Risk Society: Towards a New Modernity.* Sage Press, London.

Belzer, R. (1998) HACCP for regulatory analysis. In: *NE-165 Conference, The Economics of HACCP: New Studies of Costs and Benefits. June 1998, Washington, DC.* Also in: Unnevehr, L. (ed.) (2000) *The Economics of HACCP: New Studies of Costs and Benefits.* Eagan Press, St Paul, Minnesota. (see also www.umass.edu/ne165/haccp1998/haccp_1998.html)

Berg, P., Baltimore, D., Boyer, H.W., Cohen, S.N., Davis, R.W., Hogness, D.S., Nathans, D., Roblin, R., Watson, J.D., Weissman, S. and Zinder, N.D. (1974) Potential hazards of recombinant DNA molecules. *Science* 185.

Bergsten, C.F. (1996) Globalizing free trade. *Foreign Affairs* 75(3), 105–120.

BEUC (Bureau Europ̂een des Unions de Consommateurs) (1999) *Consumer Acceptance of Biotechnology: Will the European Parliament Get the Law Right the Second Time Round?* Press Release, BEUC 99.04, 2 February, Brussels.

Bhagwati, J.N. and Hudec, R. (eds) (1996) *Fair Trade and Harmonization.* MIT Press, Cambridge, Massachusetts, pp. 227–280.

Bode, T. (1998) Executive Director, Greenpeace International, quoted in *Economist* (1998) 348(8079), p. 76.

Bonny, S. (1999) Will biotechnology lead to more sustainable agriculture? Paper presented at the International Consortium on Agricultural Biotechnology Research (ICABR) 1999 Conference, The Shape of the Coming Agricultural Biotechnology Transformation: Strategic Investment and Policy Approaches from an Economic Perspective. University of Tor Vergata, Rome, Italy, 17–18 June 1999.

Bourdieu, P. and Wacquant, L. (1992) *An Invitation to Reflexive Sociology.* Polity Press, Oxford.

Bratton, W.W. and McCahery, J. (1996) Regulatory competition as regulatory capture: the case of corporate law in the USA. In: McCahery, J., Bratton, W.W., Picciotto, S. and Scott, C. (eds) *International Regulatory Competition and Coordination: Perspectives on Economic Regulation in Europe and the United States.* Clarendon Press, Oxford, pp. 207–238.

Bratton, W.W., McCahery, J., Picciotto, S. and Scott, C. (1996) Introduction: regulatory competition and institutional evolution. In: Bratton, W., McCahery, J., Picciotto, S. and Scott, C. (eds) *International Regulatory Competition and Coordination: Perspectives on Economic Regulation in Europe and the United States.* Clarendon Press, New York, pp. 1–55.

Brennan, G. and Buchanan, J.M. (1980) *The Power to Tax: Analytical Foundations of a Fiscal Constitution.* Cambridge University Press, Cambridge.

Brenner, K.D. (1998) *Emerging and Future Bio-engineered Commodities.* Agricultural Outlook Forum (AOF), Economic Research Service, USDA. Released 23 February.

Brill, W.J. (1988) Why engineered organisms are safe. *Issues in Science and Technology* 4(3), 44–50.

Bruhn, C.M. (1993) What consumers want to know about biotechnology. Paper presented in Workshop on Labeling of Biotechnology Products, Purdue University, Indiana, 5 June. NABC, Ithaca, New York, pp. 26–31.

Buckingham, D., Gray, R., Phillips, P., Roberts, T., Bryce, J., Morris, B., Stovin, D., Isaac, G. and Anderson, B. (1999) *The International Coordination of Regulatory Approaches to Products of Biotechnology: A Draft Final Report for Agriculture and Agri-Food Canada (AAFC).* AAFC, Ottawa.

Bureau, J.C., Marette, S. and Schiavina, A. (1997) Trade, labels and consumer information: the case of hormone-treated beef. Contributed paper at the XXII Conference of IAAE, IAAE, Sacramento, California.

CAC (Codex Alimentarius Commission) (1995) *Report of the 21st Session, Rome, 3–7 July.* ALINORM 95/37, Appendix 2, FAO, Rome.

CAC (Codex Alimentarius Commission) (1996) *Report of the Twelfth Session of the Codex Committee on General Principles, Paris, France, 23–26 November.* ALINORM 97/33, FAO, Rome.

CAC (Codex Alimentarius Commission) (1997) *Codex Alimentarius Commission Procedural Manual,* 10th edn, Articles 1(a)–(e). FAO, Rome.

CAC (Codex Alimentarius Commission) (1999) *Report of the 27th Session of the Codex Committee on Food Labelling, Ottawa, Canada, 27–30 April.* ALINORM 99/22A, FAO, Rome.

Caldwell, D.J. (1998) *The WTO Beef Hormone Ruling: an Analysis.* Community Nutrition Institute, Washington, DC.

Camerer, C. and Kunreuther, H. (1989) Decision processes for low-probability events: policy implications. *Journal of Policy Analysis and Management* 8(4), 565–592.

Cantley, M. (1995) The regulation of modern biotechnology: a historical and European perspective: a case study in how societies cope with new knowledge in the last quarter of the twentieth century. In: Rehm, H.J. and Reed, G. (eds) *Biotechnology.* Weinheim, Germany.

Cantley, M. (1998) Seeds of self-destruction: over-regulation is stifling biotechnology in Europe. In: Personal View, *Financial Times London,* December.

Cape, R.E. (1986) Future prospects in biotechnology: a challenge to United States leadership. In: Perpich, J.G. (ed.) *Biotechnology in Society: Private Initiatives and Public Oversight.* Pergamon Press, New York, pp. 5–9.

Carney, W.J. (1995) The political economy of competition for corporate charters. In: Wouters, J. and Schneider, H. (eds) *Current Issues of Cross-Border Establishment of Companies in the European Union.* Maklu, Antwerp.

Caswell, J.A. (1999) An evaluation of risk analysis as applied to agricultural biotechnology (with a case study of GMO labelling). Paper presented at the NE-167 1999 Conference Transitions in Agbiotech: Economics of Strategy and Policy, Washington, DC, 24–25 June.

Caswell, J.A. and Henson, S. (1997) Interaction of private and public food quality control systems in global markets. Conference paper presented at Globalisation of the Food Industry: Policy Implications, University of Reading, 18 and 19 September.

Caswell, J.A. and Hooker, N.H. (1996) HACCP as an international trade standard. *American Journal of Agricultural Economics* 78, 775–779.

Central Soya Protein Group (1996) Press release. 11 December.

Chess, C. (1998) The public and agricultural biotechnology. In: Hardy, R. and Segelken, J. (eds) *Agricultural Biotechnology: Novel Products and New Partnerships*. NABC Report 8, NABC, Ithaca, New York.

CI (Consumers International) (1999a) Statement of Julian Edwards, Director General of Consumers International, to the CCFL, 28 April, Ottawa. (www.oneworld.org/consumers//campaigns/food/codex/julian0499.html)

CI (Consumers International) (1999b) Comments on the proposed draft recommendations for the labelling of foods obtained through biotechnology. Presented at the 27th Session of the CCFL, 27–30 April, Ottawa. (www.oneworld.org/consumers//campaigns/foods/codex/doc.1–0499.html)

CI (Consumers International) (1999c) General surveys on foods produced through biotechnology. (www.oneworld.org/consumers//campaigns/food/codex/survey0499.html)

CIB (Canadian Institute of Biotechnology) (1997) About biotechnology: the communications experience in Canada (working draft). Contributed paper, Biotechnology and the Consumer, Conference of the Office of Consumer Affairs, Industry Canada, Ottawa.

Clark, J. (1991) *Democratizing Development: the Role of Volunteer Organizations*. Earthscan, London.

Codex Executive Committee (CEC) (1996) *Risk Analysis in Codex Work: Progress Report for Consideration of the 43rd Session of the Codex Executive Committee, Geneva, 4–7 June*. CX/EXEC 96/43/6, FAO, Rome.

Codex Secretariat (1993) *Codex Medium-Term Programme of Work 1993–98 Codex*. ALINORM 93/40 Paragraphs 75–79, App. III, FAO, Rome.

Codex Secretariat (1999) *Report of the 14th Session of the Codex Committee on General Principles, Paris, 13–19 April*. CX/GP 99/7, FAO, Rome.

Coffey, P. (1993) *The EC and the United States*. Pinter Publishers, London.

Commission on Global Governance (1995) *A Call to Action: A Summary of Our Global Neighbourhood*. Oxford University Press, Oxford. (www.cgg.ch/execsum.htm)

Connor, J. (1994) North America as a precursor of changes in Western European food purchasing patterns. *European Review of Agricultural Economics* 21, 155–173.

Consumers' Association of Canada (1994) *Statement of Policy: Regulating Agricultural Products of Biotechnology*. Consumers' Association of Canada, Ottawa.

Consumers' Association of Canada (1999) *Statement on Biotechnology*. November. Consumers' Association of Canada, Ottawa.

Consumers' Association UK (1997) *Gene Cuisine – a Consumer Agenda for Genetically Modified Foods.* Consumers' Association, London.

Consumers' Choice Council (1999) GMOs/LMOs and labeling in the context of the Biosafety Protocol negotiations. A letter to Under-Secretary Frank Loy, US Department of State, 9 February. CCC, Washington, DC. (www.consumerscouncil.org)

Cooper, C.A. and Massell, B.F. (1965) A new look at customs union theory. *Economic Journal* 75, 742–747.

Daily Telegraph (22 May 1999) GM food 'no worse than traditional plant crop'. C. Clover.

Daily Telegraph (10 February 2001) Greenpeace backs down on GM rice. R. Highfield. (www.telegraph.co.uk/e...tmo=99999999&pg=/ct/01/2/10/ngree 10.html)

Daly, H. (1971) Toward a stationary-state economy. In: Harte, J. and Socolow, R. (eds) *Patient Earth.* Holt, Rinehart and Winston, New York.

Daly, H. and Cobb, J., Jr (1994) *For the Common Good, Redirecting the Economy Toward Community, the Environment and a Sustainable Future,* 2nd edn. Beacon Press, Boston, Massachusetts.

David, P.A. (1987) Some new standards for the economics of standardization in the information age. In: Dasgupta, P. and Stoneman, P. (eds) *Economic Policy and Technological Performance.* Cambridge University Press, Cambridge.

Dezalay, Y. (1996) Between the state, law and the market: the social and professional stakes in the construction and definition of a regulatory arena. In: Bratton, W., McCahery, J., Picciotto, S. and Scott, C. (eds) *International Regulatory Competition and Coordination: Perspectives on Economic Regulation in Europe and the United States.* Clarendon Press, Oxford, pp. 59–88.

Dickman, S. (1989) Europe avoids moratorium. *Nature* 339, 413.

Drache, D. (1996) Dreaming trade or trading dreams: the limits of trade blocs. In: Bratton, W., McCahery, J., Picciotto, S. and Scott, C. (eds) *International Regulatory Competition and Coordination: Perspectives on Economic Regulation in Europe and the United States.* Clarendon Press, Oxford, pp. 417–443.

Ecologist (1998) The Monsanto files: can we survive genetic engineering? 28(5), September/October.

Economist (1996) A moral maize. 14 September, p. 82.

Economist (June 1998) In defence of the demon seed. 13 June, p. 13.

Economist (June 1999a) Food for thought. 351(8124), 19 June, pp. 23–25.

Economist (1999b) Why greens should love trade. 353(8140), 9 October, p. 18.

Economist (1999c) Embracing greenery. 353(8140), 9 October, pp. 125–126.

Einseidel, E.F. (1997) Biotechnology and the Canadian public: report on a 1997 national survey and some international comparisons. Contributed paper, Biotechnology and the Consumer, Conference of the Office of Consumer Affairs. Industry Canada, Ottawa.

Enders, A. (1997) *Openness and the WTO: a Draft IISD Working Paper.* International Institute for Sustainable Development, Winnipeg. (iisd1.iisd.ca/)

EPA (Environmental Protection Agency), United States (1992) *EPA Proposal to Clarify the Regulatory Status of Plant Pesticides.* Federal Register 57:55531,Washington, DC.

Ernst and Young (1995) *Regulatory Case Studies: Final Report.* Research Paper for Western Economic Diversification.

EU Food Law Monthly (1999) No. 95, November. AgraEurope, London.

Eurobaromètre (Biotechnology and the European Public Concerted Action Group) (1997) Europe ambivalent of biotechnology: a commentary. *Nature* 387, 845–847.

EuroFood (1999) *Slowly Moving Towards a European Food Safety Agency.* AgraEurope, London.

EuropaBio (1997) *Biotechnology: Consumer/User Information and Informed Choice for Products of GM and Non-GM Sources.* Position Document No. 267, June, Brussels.

European Commission (1983) *Biotechnology in the Community, Communication from the Commission to the Council.* COM(83)672 Final, Brussels. 4 October.

European Commission (1985) *White Paper on Completing the Internal Market* COM(85) 310 Final, Brussels.

European Commission (1986) *A Community Framework for the Regulation of Biotechnology, Communication from the Commission to the Council.* COM(86)573 Final, Brussels. 4 November.

European Commission (1988a) *The European Community and the Contained Use of Genetically Modified Micro-Organisms.* DG XI/A/2, Brussels.

European Commission (1988b) *Proposal for a Council Directive on the Deliberate Release to the Environment of Genetically Modified Organisms* COM(88)160 Final – SYN 131, Brussels. *Official Journal* C 198, 28 July, 19–27.

European Commission (1991) *Commission Communication to Parliament and the Council: Promoting a Competitive Environment for Industrial Activities Based on Biotechnology Within the Community.* Brussels.

European Commission (1994a) *Growth, Competitiveness and Employment: The Challenges and Ways Forward into the 21st Century.* White Paper, Luxembourg.

European Commission (1994b) *The Views of the Research Community on the Regulatory Framework: DG XII Survey, Executive Summary.* Brussels.

European Commission (1998) *Commission Decision of 22 April 1998 concerning the placing on the market of genetically modified spring swede rape* (Brassica napus *L. ssp.* oleifera), *pursuant to Council Directive 90/220/EEC.* 98/291/EC, Brussels. *Official Journal* L No. 131, 5 May.

European Commission (1999) *The Millennium Round: an Economic Appraisal. Directorate-General for Economic and Financial Affairs.* November. ECFin/659/99 – Rev.EN, Brussels.

European Council (1990a) *Council Directive of 23 April 1990 on the Contained Use of Genetically Modified Micro-organisms.* 90/219/EEC, Brussels. *Official Journal* L No. 117, 8 May.

European Council (1990b) *Council Directive of 23 April 1990 on the Deliberate Release into the Environment of Genetically Modified Organisms.* 90/220/EEC, Brussels. *Official Journal* L No. 117, 8 May.

European Federation of Biotechnology (1998) *Lessons from the Swiss Biotechnology Referendum.* Briefing Paper No. 8.

European Parliament (1987) *Resolution on Biotechnology in Europe and the Need for an Integrated Policy.* Doc A2–134/86, Brussels. *Official Journal* C 76, 23 March.

European Parliament (1997) *Novel Foods Regulation 258/97.* Brussels.

European Parliament (1999) *Opinion on a Draft Commission Regulation (EC) Amending Council Regulation EC No. 1139/98 Concerning the Compulsory Indication on the Labelling of Certain Foodstuffs Produced from Genetically Modified Organisms of Particulars Other than Those Provided for in Directive 79/112/EEC.* Brussels. February.

Europe Drug and Device Letter (1996) 5 February, p. 7.

Evans, P. (1999) Draining the swamp: votes and tariffs in the age of globalisation. *Consumer Policy Review* 9(3), 104–110.

Falconi, C. (1999) Agricultural biotechnology research indicators in developing countries. Paper presented at the International Consortium on Agricultural Biotechnology Research (ICABR) 1999 Conference, The Shape of the Coming Agricultural Biotechnology Transformation: Strategic Investment and Policy Approaches from an Economic Perspective, University of Tor Vergata, Rome, Italy, 17–18 June 1999.

FAO (Food and Agricultural Organisation) (1998) Understanding the Codex Alimentarius: origins of the Codex Alimentarius. (www.fao.org/docrep/w9114e/W9114e01htm)

FAO/WHO (Food and Agricultural Organization/World Health Organization) (1950) *Report of the First Meeting of the Joint FAO/WHO Expert Committee on Nutrition.*

FAO/WHO (Food and Agricultural Organization/World Health Organization) (1995) *Expert Consultation on the Application of Risk Analysis to Food Standards Issues.* WHO/FNU/FOS 95.3, 13–17 March, Rome.

FDA (Food and Drug Administration), United States Department of Health and Human Services (1992) *Statement of Policy: Foods Derived From New Plant Varieties.* Federal Register 29 May 1992 57:22984, Washington, DC.

Financial Times (1999a) Monsanto admits promotion of modified foods has backfired. 15 March.

Financial Times (1999b) Supermarkets in move on modified foods: J. Willman. 18 March.

Financial Times (1999c) Germans catching up on UK biotechnology groups: D. Pilling. 20 April.

Financial Times (1999d) Genetically modified crops feed campaigners' anger: D. Pilling. 22 May.

Financial Times (1999e) The juggernaut and the butterfly: C. Cookson. 22 May.

Financial Times (1999f) GM foods declared safe by government scientists: V. Houlder and G. Parker. 22 May.

Financial Times (1999g) Poverty 'compels' GM crop research: C. Cookson. 28 May.

Financial Times (1999h) France seeking suspension in sales of modified foods: R. Graham. 24 June.

Financial Times (1999i) Rules on genetically modified food linger in political limbo: M. Smith. 24 June.

Financial Times (1999j) EU unlikely to back new modified foods: M. Smith. 25 June.

Financial Times (1999k) Rules for labelling GM foods 'will need new cash': V. Houlder. 25 June.
Fincham, J.R.S. and Ravetz, J.R. (1991) *Genetically Engineered Organisms: Benefits and Risks.* The Open University Press in collaboration with the Working Party of the Council for Science and Society, Milton Keynes.
Foster, K.R., Vecchia, P. and Repacholi, M.H. (2000) Science and the precautionary principle. *Science* 12, 979–981.
Fox, M.W. (1990) Transgenic animals: ethical and welfare concerns. In: Wheale, P. and McNally, R. (eds) *The BioRevolution: Cornucopia or Pandora's Box?* Pluto Press, London, pp. 31–45.
Fox, M.W. (1992) *Superpigs and Wondercorn: the Brave New World of Biotechnology … and Where It May All Lead.* Lyons and Burford, New York.
Frawley, J.P. (1987) Codex Alimentarius – food safety – pesticides. *Food Drug Cosmetic L.J.* 42, 168.
Frey, B. (1984) *International Political Economics.* Martin Robertson, Oxford.
Frey, B. (1985) The political economy of protection. In: Greenaway, D. (ed.) *Current Issues in International Trade, Theory and Policy.* Macmillan, London.
Frost, E. (1998) A transatlantic free trade agenda? Prospects for transatlantic cooperation in shaping the future multilateral trading order. In: van Scherpenberg, J. and Thiel, E. (eds) *Towards Rival Regionalism? US and EU Regional Regulatory Regime Building.* Nomos Verlagsgesellschaft, Baden-Baden, pp. 277–303.
Fuchs, R.L. (1993) Risk assessment: a technical perspective. In: Fessenden MacDonald, J. (ed.) *AgriculturalBiotechnology: a Public Conversation About Risk.* NABC Report 5, NABC, Ithaca, New York, pp. 55–64.
GAFTA (Grain and Free Trade Association) (1997) Position paper to Sir Leon Brittan. 28 May.
Garvey, J.I. (1995) Trade law and the quality of life – dispute resolution under the NAFTA side accords on labor and the environment. *American Journal of International Law* 89, 439–453.
Gaskell, G. (2000) Agricultural biotechnology and public attitudes in the European Union. *AgBioForum* 3(23). (www.agbioforum.org)
Gasson, M.J. (1999) Genetically modified foods face rigorous safety evaluation. *Nature* 402, 229.
Gaster, R. and Prestowitz, C. (1994) *Shrinking the Atlantic: Europe and the American Economy.* North Atlantic Research, Washington, DC.
Gatsios, K. and Holmes, P. (1998) Regulatory competition. In: Newman, P. (ed.) *New Palgrave Dictionary of Economics and the Law.* Palgrave/St Martin's Press, Virginia.
GATT (General Agreement on Tariffs and Trade) (1994) *Multilateral Negotiations/The Uruguay Round: Trade Negotiations Committee, Final Act Embodying the Results of the Uruguay Round of Multilateral Trade Negotiations.* GATT Secretariat, Geneva.
Giannakas, K. and Fulton, M. (2000) Market effects of genetic modification under alternative labeling regimes and enforcement scenarios. Paper presented at the 2000 American Agricultural Economics Association Annual Meeting, Tampa, Florida.
Giddens, A. (1994) Living in a post-traditional society. In: Beck, U., Giddens, A. and Lash, S. (eds) *Reflexive Modernism.* Polity, Cambridge.

Gilpin, R. (1987) *The Political Economy of International Relations*. Princeton University Press, Princeton, New Jersey.
Globe and Mail (1998) Chemical companies bet on the farm: R. Shapiro. 27 February.
Grace, E.S. (1997) *Biotechnology Unzipped: Promises and Realities*. Trifolium Books, Toronto.
Graff, G.D., Rausser, G.C. and Small, A.A. (1999) Agricultural biotechnology's complementary intellectual assets. Paper presented at the International Consortium on Agricultural Biotechnology Research (ICABR) 1999 Conference, The Shape of the Coming Agricultural Biotechnology Transformation: Strategic Investment and Policy Approaches from an Economic Perspective, University of Tor Vergata, Rome, Italy, 17–18 June 1999.
Greene, W.H. (1997) *Econometric Analysis*, 2nd edn. Prentice Hall, Englewood Cliffs, New Jersey.
Greenpeace Business (1999) Monsanto admits 'more questions than answers' at Greenpeace Business Conference. No. 52, Dec. 1999/Jan. 2000. Greenpeace, London. (www.greenpeace.org.uk)
Grimwade, N. (1996) *International Trade Policy: a Contemporary Analysis*. Routledge, London.
Grossman, M. and Helpman, E. (1991) *Innovation and Growth in the Global Economy*. MIT Press, Cambridge, Massachusetts.
Grove-White, R. (1997) Environment, risk and democracy. In: Jacobs, M. (ed.) *Greening the Millennium*. Blackwell, Oxford.
Grove-White, R., Macnaghten, P., Mayer, S. and Wynne, B. (1997) *Uncertain World: Genetically Modified Organisms, Food and the Public Attitudes in Britain*. Centre for the Study of Environmental Change, Lancaster University, Lancaster, UK.
Guardian (1999a) Goliath sets farm gene trap: A. Simms. 10 May.
Guardian (1999b) GM giants 'will force the world into famine': J. Vidal. 10 May.
Haas, P.M. *et al.* (1992) Knowledge, power and international policy coordination. *International Organization (Special Issue)* 46(1).
Habermas, J. (1971) *Knowledge and Human Interests*. Beacon Press, Boston, Massachusetts.
Haerlin, B. (1990) Genetic engineering in Europe. In: Wheale, P. and McNally, R. (eds) *The Bio Revolution: Cornucopia or Pandora's Box?* Pluto Press, London, pp. 253–261.
Hallman, W.K. and Metcalfe, J. (1993) *Public Perceptions of Agricultural Biotechnology: A Survey of New Jersey Residents*. Rutgers University Ecosystem Policy Research Center, New Brunswick, New Jersey.
Haniotis, T. (2000) Regulating agri-food production in the US and the EU. *AgBioForum* 3(23). (www.agbioforum.org)
Hanley, B. and Johnson, M.-L. (1999) Presentation at IPP Conference, London, 2–3 December.
Harlander, S.K. (1993) Unique characteristics of biotechnology products. Paper presented in Workshop on Labeling of Biotechnology Products. Purdue University, Indiana, 5 June. NABC, Ithaca, New York, pp. 16–25.
Hausman, D.M. (1998) Problems with realism in economics. *Economics and Philosophy* 14, 185–213.

Hay, J. (2000) *Practising Democratic Foreign Policy: DFAIT's Consultations with Canadians*. A Report to the Canadian Centre for Foreign Policy Development.

Health Canada (1994) *Guidelines for the Assessment of Novel Foods*. Health Canada, Government of Canada, Ottawa.

Ho, M.-W. (1998) *Genetic Engineering: Dream or Nightmare? The Brave New World of Bad Science and Big Business*. Gateway Books, Bath, UK.

Hoban, T.J. (1996a) Trends in consumer attitudes about biotechnology. *Journal of Food Distribution* 27, 1–10.

Hoban, T.J. (1996b) How Japanese consumers view biotechnology. *Journal of Food Technology* 50, 85–88.

Hoban, T.J. (1996c) Consumers will accept biotech foods. *Biotech Catalyst*.

Hoban, T.J. (1997) Consumer acceptance of biotechnology: an international perspective. *Nature* 15(3), 232–234.

Hoban, T.J. and Kendall (1992) *Consumer Attitudes About the Use of Biotechnology in Agriculture and Food Production*. Report to the USDA Extension Service.

Holloway, G.J. (1999) Evaluating the alternatives. *American Journal of Agricultural Economics* 81(5), 1090–1095.

Horton, L. (1997) International harmonisation of food and veterinary medicine regualtion. In: *Fundamentals of Law and Regulation*, pp. 381–483.

Houston, D.L. (1987) Codex Alimentarius Commission: 25 years of fair trade and consumer protection. *Food Drug Cosmetic L.J.* 42, 163–164.

Hullar, T.L. (1993) Putting it in context. In: Fessenden MacDonald, J. (ed.) *Agricultural Biotechnology: a Public Conversation About Risk*. NABC Report 5, NABC, Ithaca, New York, pp. 13–18.

Independent (1999a) Tesco isolated over policy on GM food: P. Waugh and M. McCarthy. 18 March.

Independent (1999b) Victory for grass roots action: G. Lean. 2 May.

Independent (1999c) Let the public act as official GM crops monitors, say MPs: P. Waugh. 14 May.

Independent (1999d) A science lesson that we must all digest. 19 May.

Independent (1999e) Royal Society delivers damning verdict on GM foods research: S. Connor. 19 May.

Independent (1999f) Blair wrong on GM crops, says Chief Government Scientist: P. Waugh and C. Arthur. 20 May.

Independent (1999g) When science facts become science fiction: S. Connor. 22 May.

Independent (1999h) Blair fumes, but public backs Charles' stand against GM: R. Sylvester, M. Woolf and G. Lean. 6 June.

Independent (1999i) It's nature's way to talk out of turn: J. Smith. 6 June.

Independent (1999j) Cherie's against GM too, says the Prince. 6 June.

Independent (1999k) UK rejects ban on GM crops: S. Castle and C. Brown. 25 June.

Independent (1999l) GM technology to restore lost elm: M. Rowe. 8 August.

Independent (1999m) Marks goes squeaky clean: J. Blythman. 5 September.

Independent (1999n) Iceland discovers that there's profit in purity: D. Gledhill. 5 September.

Independent (1999o) Blair think-tank calls for EU referendum revolution: S. Castle. 5 September.

Independent (1999p) Garden writers gang up on GM: M. Rowe. 5 September.
Independent (1999q) GM bosses want to pull out of UK: G. Lean. 5 September.
Independent (1999r) Humbling of a GM giant: G. Lean. 3 October.
Independent (1999s) Smeared GM expert vindicated: G. Lean. 30 October.
Independent (1999t) GM Foods: the Debate, A Special Report Produced in Association with Iceland. 12 October.
Industry Canada (1997) *Canadian Biotechnology Strategy.* Industry Canada, Ottawa.
Ingleheart, R. (1997) *Modernization and Post-Modernization: Cultural, Economic and Political Change in 43 Societies.* Princeton University Press, Princeton, New Jersey.
International Institute for Sustainable Development (1994) *Trade and Sustainable Development Principles (the 'Winnipeg Principles').* IISD, Winnipeg.
Isaac, G.E. (2001) *International Trade and Citizen Engagement: Transparency and Participation in the Trade Policy Advisory System.* Final Research Report for the Canadian Department of Foreign Affairs and International Trade, Estey Centre for Law and Economics in International Trade, Saskatoon, Canada.
Isaac, G.E. and Phillips, P.W.B. (1999a) *Assessment of the Potential Impact of the Biosafety Protocol: the Canadian Agricultural Commodities Case.* Research Paper Commissioned by the Canadian Food Inspection Agency, Final Report submitted 1 March 1999.
Isaac, G.E. and Phillips, P.W.B. (1999b) Market access and market acceptance for genetically modified products. Paper presented at the International Consortium on Agricultural Biotechnology Research (ICABR) 1999 Conference, The Shape of the Coming Agricultural Biotechnology Transformation: Strategic Investment and Policy Approaches from an Economic Perspective, University of Tor Vergata, Rome, Italy, 17–18 June 1999.
Isaac, G.E. and Phillips, P.W.B. (1999c) The Biosafety Protocol and international trade in transgenic canola: an economic assessment of the impact on Canada. Paper presented at the NE-167 1999 Conference, Transitions in Agbiotech: Economics of Strategy and Policy, Washington, DC, 24–25 June.
Isaac, G.E. and Phillips, P.W.B. (2001) Regulating domestic markets. In: Phillips, P.W.B. and Khachatourians, G.G. (eds) *The Biotechnology Revolution in Global Agriculture: Invention, Innovation and Investment in the Canola Sector.* CAB International, Wallingford, UK, pp. 213–241.
Isaac, G.E. and Woolcock, S.B. (1999a) *Green Labels: Consumer Interests and Transatlantic Trade Tensions in Ecolabelling.* Research Project for the Consumers International Project, Support to Consumer Organisations in Promoting Sustainable Consumption.
Isaac, G.E. and Woolcock, S.B. (1999b) *Food Safety and Trade Policy: Agricultural Biotechnology Issues.* Consumers' Association, London.
Jackson, J. (1996) WTO dispute settlement procedures: a preliminary assessment. In: Schott, J. (ed.) *The World Trading System: Challenges Ahead.* Institute for International Economics, Washington, DC, pp. 153–165.
Jackson, L.A. (1999) Agricultural standard harmonization in international

trading systems. Paper presented at the International Consortium on Agricultural Biotechnology Research (ICABR) 1999 Conference, The Shape of the Coming Agricultural Biotechnology Transformation: Strategic Investment and Policy Approaches from an Economic Perspective, University of Tor Vergata, Rome, Italy, 17–18 June 1999.

James, C. (1997) *Global Status of Transgenic Crops in 1997*. International Service for the Acquisition of AgBiotech Applications (ISAAA) Briefs Series No. 5, ISAAA, Ithaca, New York.

James, C. (1999) Transgenic crops worldwide: current situation and future outlook. Paper presented at the Conference, Agricultural Biotechnology in Developing Countries: Towards Optimizing the Benefits for the Poor, Organized by ZEF and ISAAA in Collaboration with AgrEvo and DSE in Bonn, 15–16 November 1999.

James, C. (2000) *Global Review of Commercialized Transgenic Crops*. International Service for the Acquisition of AgBiotech Applications (ISAAA) Briefs Series, ISAAA, Ithaca, New York.

Johnson, H.G. (1965) An economic theory of protectionism, tariff bargaining, and the formation of customs unions. *Journal of Political Economy* 73, 256–283.

Johnson, P.M. (2000) *Beyond Trade: the Case for a Broadened International Governance Agenda*. Policy Matters 1(3), Institute for Research on Public Policy, Ottawa.

Joly, P.B. and Lemarie, S. (1998) Industry consolidation, public attitude and the future of plant biotechnology in Europe. *AgBioForum* 1(2), 85–90.

Jovanovic, M.N. (ed.) (1998a) *International Economic Integration: Critical Perspectives on the World Economy* – Vol. I. *Theory and Measurement*. Routledge, London.

Jovanovic, M.N. (ed.) (1998b) *International Economic Integration: Critical Perspectives on the World Economy* – Vol. II. *Monetary, Fiscal and Factor Mobility Issues*. Routledge, London.

Jovanovic, M.N. (ed.) (1998c) *International Economic Integration: Critical Perspectives on the World Economy* – Vol. III. *General Issues*. Routledge, London.

Jovanovic, M.N. (ed.) (1998d) *International Economic Integration: Critical Perspectives on the World Economy* – Vol. IV. *Integration Schemes*. Routledge, London.

Jovanovic, M.N. (1998e) *International Economic Integration: Limits and Prospects*. Routledge, London.

Kalaitzandonakes, N. (2000) Why does biotech regulation differ so much between the US and the EU? *AgBioForum* 3(23). (www.agbioforum.org)

Kalaitzandonakes, N. and Hayenga, M. (1999) Structural change in the biotechnology and seed industrial complex: theory and evidence. Paper presented at the NE-167 1999 Conference, Transitions in Agbiotech: Economics of Strategy and Policy, Washington, DC, 24–25 June.

Katz, M. and Shapiro, C. (1986) Technology adaptation in the presence of network externalities. *Journal of Political Economy* 94, 822–841.

Kearns, P. and Mayers, P. (1999) Substantial equivalence is a useful tool. *Nature* 402, 640.

Keck, M. and Sikkink, K. (1994) Transnational issue networks in international politics. Unpublished conference paper.

Kenney, M. (1986) *Biotechnology: The University–Industrial Complex.* Yale University Press, New Haven, Connecticut.

Kneen, B. (1999) *Farmageddon: Food and the Culture of Biotechnology.* New Society Publishers, British Columbia.

Kobrin, S.J. (1998) The MAI and the clash of globalization. *Foreign Policy* 1998 (Autumn), 97–109.

Korwek, E.L. (1993) Current status of food labels. Paper presented in Workshop on Labeling of Biotechnology Products, Purdue University, Indiana, 5 June. NABC, Ithaca, New York, pp. 3–15.

Koster, M.J. and Balf, T.P. (1998) Mandatory and permissive labelling of genetically engineered food products. *Journal of Biolaw and Business* 1(3), 65–69.

Kramer, C.S. (1989) Food safety and international trade: the US–EC meat hormone controversies. In: Kramer, C.S. (ed.) *The Political Economy of US Agriculture: Challenges for the 1990s.* Annual Policy Review – Resources for the Future, National Center for Food and Agricultural Policy, Washington, DC, pp. 203–233.

Kraus, M. (1998) Innovation, industrial development and the regulation of biotechnology. In: Hardy, R. and Segelken, J. (eds) *Agricultural Biotechnology: Novel Products and New Partnerships.* NABC Report 8, NABC, Ithaca, New York.

Krimsky, S. and Wrubel, R. (1996) *Agricultural Biotechnology and the Environment.* University of Illinois Press, Urbana, Illinois.

Landfried, C. (1997) Beyond technocratic governance: the case of biotechnology. *European Law Journal* 3(3), 255–272.

Langille, B. (1996) Competing conceptions of regulatory competition in debates on trade liberalization and labour standards. In: Bratton, W., McCahery, J., Picciotto, S. and Scott, C. (eds) *International Regulatory Competition and Coordination: Perspectives on Economic Regulation in Europe and the United States.* Clarendon Press, Oxford, pp. 479–490.

Lavoie, B.F. and Sheldon, I.M. (1999) The source of comparative advantage: a real options approach. Paper presented at the NE-167 1999 Conference, Transitions in Agbiotech: Economics of Strategy and Policy, Washington, DC, 24–25 June.

Lipschutz, R.D. (1992) Reconstructing world politics: the emergence of global civil society. *Millennium* 21(3), 389–420.

Lipsey, R.G. (1988) Sovereignty: culturally, economically and socially. In: Crispo, J. (ed.) *Free Trade – The Real Story.* Gage, Toronto, pp. 148–160.

Lister, C. (1992) Protectionism and integration: designations of origin for foodstuffs in the European Community. *Food and Drug* 47, 641–644.

Lister, C. (1996) A sad story told sadly: the prospects for US–EU food trade wars. *Food and Drug* 51, 303.

McCormick, J. (1998) Interest groups and the media in post-Cold War U.S. foreign policy. In: Scott, J.M. (ed.) *After the End: Making U.S. Foreign Policy in the Post-Cold War World.* Duke University Press, Durham, North Carolina, pp. 171–198.

McDonald (1997) *An Analysis of the Potential Impacts to the Canadian Biotechnology Sector and Canada's Export of Agricultural and Other Commodities of the Proposed International Biosafety Protocol.* Research Paper for AgWest Biotech, Saskatoon, Canada.

McDonald, J. (1998) Big beef up or consumer health threat?: The WTO food safety agreements, bovine growth hormone and the precautionary principle. *Environmental Planning and Law Journal* 15(2), 115–123.

McHughen, A. (2000) *Pandora's Picnic Basket: the Potential Hazards of Genetically Modified Foods.* Oxford University Press, New York.

MacKenzie, D.R. (1993) Regulatory risk assessment: a view from the Potomac. In: Fesseden MacDonald, J. (ed.) *Agricultural Biotechnology: a Public Conversation About Risk.* NABC Report 5, NABC, Ithaca, New York, pp. 47–54.

MacLaren, D. (1997) Uncertainty aversion and technical barriers to trade: an Australian example. In: Orden, D. and Roberts, D. (eds) *Understanding Technical Barriers to Agricultural Trade: Proceedings of a Conference of the International Agricultural Trade Research Consortium (IATRC).* IATRC, St Paul, Minnesota, pp. 225–283.

Madden, D. (1995) *Briefing Notes from the First UK Consensus Conference on Plant and Microbial Biotechnology, 1994.* National Centre for Biotechnology Education, University of Reading, Reading.

Mail on Sunday (1999) Up with Prince Charles. Letter to the Editor: V. Pride.

Majewski, C.M. (1997) *Food Safety Strategies in the Changing Global Environment.* Report of the 1996–97 Government Study Fellowship of the European Commission June 1997, Brussels.

Majone, G. (ed.) (1990) *Deregulation vs. Reregulation: Regulatory Reform in Europe and the United States.* Pinter Publishers, London.

Majone, G. (1993) Comparing strategies for regulatory rapprochement. Contributed paper, OECD Symposium on Managing Regulatory Regimes Between Different Levels of Government. Organisation for Economic Cooperation and Development (OECD), Paris.

Majone, G. (1994) *Interdependence vs. Accountability: Non-Majoritarian Institutions and Democratic Government in Europe.* EUI Working Paper SPS 94/3, European University Institute, San Domenico di Fiesole.

Malchup, F. (1958) *An Economic Review of the Patent System.* Study No. 15. A Study of the Subcommittee on Patents, Trademarks and Copyright, Committee of the Judiciary, US Senate, Washington, DC.

Malchup, F. (1979) *A History of the Thought on Economic Integration.* Macmillan, London.

Manitoba Pool Elevators (1997) Identity preserved canola production in Canada, 1996. Discussion paper.

Mayer, S., Hill, J., Grove-White, R. and Wynne, B. (1996) *Uncertainty, Precaution and Decision-Making: the Release of Genetically Modified Organisms into the Environment.* UK ESRC Global Environmental Change Programme Briefings, No. 8, 8 June.

Metro (8 June 1999) GM farmer: destroying crops was a mistake.

Milward, A.S. (1992) *The European Rescue of the Nation-State.* Routledge, London.

Moschini, G., Lapan, H. and Sobolevsky, A. (1999) Trading technology as well as final products: Roundup Ready soybeans and welfare effects in the soybean complex. Paper presented at the International Consortium on Agricultural Biotechnology Research (ICABR) 1999 Conference, The Shape of the Coming Agricultural Biotechnology Transformation: Strategic Investment and Policy Approaches from an Economic Perspective, University of Tor Vergata, Rome, Italy, 17–18 June 1999.

Mulongoy, K. (1997) Different perspectives on the International Biosafety Protocol. *Biotechnology and Development Monitor* 30. The Hague.

Najam, A. (1996) NGO accountability: a conceptual framework. *Development Policy Review* 14(4), 339–354.

National Academy of Sciences (NAS) Committee on the Institutional Means for Assessment of Risks to Public Health, Commission of Life Sciences (1983) *Risk Assessment in the Federal Government: Managing the Process.* National Research Council, National Academy Press, Washington, DC.

National Consumers Council (1998) *Farm Policies and Our Food: The Need for Change.* PD 11/B2/98, NCC, London.

National Soybean Processors Association (1996a) Position paper to Mr Wood, Food and Drink Federation (UK), 31 October.

National Soybean Processors Association (1996b) Position paper to Mr Vigeveno, Unilever, 2 December.

Nature Biotechnology (1996) Ciba faces a maize of committees in Europe. *Nature Biotechnology* 14, 1068.

Nuffield Council on Bioethics (1999) *Genetically Modified Crops: the Ethical and Social Issues.* Nuffield Council, London.

OECD (Organization for Economic Cooperation and Development) (1993) *Environmental Policies and Industrial Competitiveness.* OECD, Paris.

OECD (Organization for Economic Cooperation and Development) (1997a) *The Knowledge-based Economy.* OECD, Paris.

OECD (Organization for Economic Cooperation and Development) (1997b) *Report on Regulatory Reform: Summary.* OECD, Paris.

OECD (Organization for Economic Cooperation and Development) (1997c) *Processes and Production Methods (PPMs): Conceptual Framework and Considerations on the Use of PPM-Based Trade Measures.* OECD/GD(97)/37, Paris.

Office of Technology Assessment (1984) *Commercial Biotechnology: An International Analysis.* US Government Printing Office, Washington, DC.

Office of Technology Assessment (1986) *Coordinated Framework of Regulation of Biotechnology: Announcement of Policy Notice for Public Comment.* Federal Register 51 233OR, 23350, 26 June.

Office of Technology Assessment (1991) *Biotechnology in a Global Economy.* OTA-BA-494, US Government Printing Office, Washington, DC.

Olson, M. (1965) *The Logic of Collective Action: Public Goods and the Theory of Groups.* Harvard University Press, London.

Patterson, L.A. (2000) Biotechnology policy in the European Union. In: Wallace, H. and Wallace, W. (eds) *Policy Making in the European Union*, 4th edn. Oxford University Press, Oxford.

Paul, J.R. (1996) Competitive and non-competitive regulatory markets: the regulation of packaging waste in the EU. In: McCahery, J., Bratton, W.W., Picciotto, S. and Scott, C. (eds) *International Regulatory Competition and Coordination: Perspectives on Economic Regulation in Europe and the United States.* Clarendon Press, Oxford, pp. 354–379.

Pearce, D.W. (1996) *The MIT Dictionary of Modern Economics*, 4th edn. MIT Press, Cambridge, Massachusetts.

Perdikis, N., Kerr, W.A. and Hobbs, J.E. (1999) Can WTO/GATT Agreements on Sanitary and Phyto-Sanitary Measures and Technical Barriers to Trade be renegotiated to accommodate agricultural biotechnology? Paper presented at the NE-167 1999 Conference, Transitions in Agbiotech: Economics of Strategy and Policy, Washington, DC, 24–25 June.

Phillips, P.W.B. (1991) *Wheat, Europe and the GATT.* Pinter Press, London.

Phillips, P.W.B. and Isaac, G.E. (1998) GMO labeling: threat or opportunity? *AgBioForum* 1(1). (www.agbioforum.missouri.edn)

Phillips, P.W.B. and Isaac, G.E. (2001) Regulating International trade in knowledge-based products. In: Phillips, P.W.B. and Khachatourians, G.G. (eds) *The Biotechnology Revolution in Global Agriculture: Invention, Innovation and Investment in the Canola Sector.* CAB International, Wallingford, UK, pp. 243–269.

Phillips, P.W.B. and Khachatourians, G.G. (2001) Approaches to and measurement of innovation. In: Phillips, P.W.B. and Khachatourians, G.G. (eds) *The Biotechnology Revolution in Global Agriculture: Invention, Innovation and Investment in the Canola Sector.* CAB International, Wallingford, UK, pp. 23–32.

Picciotto, S. (1996) The regulatory criss-cross: interaction between jurisdictions and the construction of global regulatory networks. In: McCahery, J., Bratton, W.W., Picciotto, S. and Scott, C. (eds) *International Regulatory Competition and Coordination: Perspectives on Economic Regulation in Europe and the United States.* Clarendon Press, Oxford, pp. 89–123.

Porter, M. (1990) *The Competitive Advantage of Nations.* The Free Press, New York.

Prakash, C.S. (1999) Comments. Who is playing God? *AgraFood Biotech* No. 20, 22 December. AgraEurope, London.

Preeg, E.H. (1996) Policy forum: transatlantic free trade. *Washington Quarterly* 19(2) (Spring), 103–133.

Public Ledger (1999a) In brief. No. 72, 139. AgraEurope, London.

Public Ledger (1999b) EC presents two opposing cases. No. 72, 139. AgraEurope, London.

Purchase, B. (1997) Rapporteur's remarks: Symposium on Biotechnology and the Consumer. Contributed paper, Biotechnology and the Consumer Conference of the Office of Consumer Affairs, Industry Canada, Ottawa.

Qaim, M. (1999) Modern Biotechnology for small-scale farmers in developing countries: contradiction or promising option? Paper presented at the International Consortium on Agricultural Biotechnology Research (ICABR) 1999 Conference, The Shape of the Coming Agricultural Biotechnology Transformation: Strategic Investment and Policy Approaches from an Economic Perspective, University of Tor Vergata, Rome, Italy, 17–18 June 1999.

Rausser, G., Scotchmer, S. and Simon, L. (1999) Intellectual property and market structure in agriculture. Paper presented at the International Consortium on Agricultural Biotechnology Research (ICABR) 1999 Conference, The Shape of the Coming Agricultural Biotechnology Transformation: Strategic Investment and Policy Approaches from an Economic Perspective, University of Tor Vergata, Rome, Italy, 17–18 June 1999.

Rayner, S. (1992) Cultural theory and risk analysis. In: Krimsky, S. and Golding, D. (eds) *Social Theories of Risk*. Praeger, Westport, pp. 83–116.

Rhodes, M. (1991) The social dimension of the European single market: national versus transnational regulation. *European Journal of Political Research* 19, 245–280.

Rhodes, M. (1994) 'Subversive liberalism': market integration, globalization and the European welfare state. Paper presented at the 23rd Annual European Consortium for Political Research, Madrid, 17–22 April.

Rifkin, J. (1998a) The second genesis: science may be ready to create a perfect world – but who will define perfect? In: Essays on the Millennium Series. *MacLeans* (Canadian Newsmagazine), 4 May, pp. 48–52.

Rifkin, J. (1998b) *The Biotech Century: Harnessing the Gene and Remaking the World*. Pitnam, New York.

Roberts, D. (1998) A classification of technical trade barriers in agricultural markets. Conference paper presented at Food Quality and Safety Standards and Trade in Agricultural and Food Products Conference. Agricultural Economic Society, London, 15 December.

Roberts, D., Josling, T. and Orden, D. (1999) *A Framework for Analyzing Technical Trade Barriers in Agricultural Markets*. Technical Bulletin No. 1876, Market and Economics Division, Economic Research Service, US Department of Agriculture.

Rogers, E.L. (1990) Environmental legislation in the USA. In: Wheale, P. and McNally, R. (eds) *The BioRevolution: Cornucopia or Pandora's Box?* Pluto Press, London, pp. 143–149.

Romano, R. (1996) Explaining American exceptionalism in corporate law. In: McCahery, J., Bratton, W.W., Picciotto, S. and Scott, C. (eds) *International Regulatory Competition and Coordination: Perspectives on Economic Regulation in Europe and the United States*. Clarendon Press, Oxford, pp. 127–152.

Rural Advancement Fund International (1998) *Terminator Trends: the 'Silent Spring' of Farmers' Rights; Seed Saving, the Public Sector and Terminator Technology*. Occasional Paper Series 5(1), Rural Advancement Fund International, Winnipeg, Canada.

Sage, G. (1999) Presentation at IPP Conference, London, 2–3 December.

Sahai, S. (1997) The bogus debate of bioethics. *Biotechnology and Development Monitor* 30, 24.

Salamon, L.M. (1994) The rise of the nonprofit sector. *Foreign Affairs* 73(4), 109–22.

Salamon, L.M. (1995) *Partners in Public Service: Government–Nonprofit Relations in the Modern Welfare State*. The Johns Hopkins University Press, Baltimore, Maryland.

Samuelson, P. (1954) The pure theory of public expenditure. *Review of Economics and Statistics* 36, 387–389.

Sands, P. (1995) *Principles of International Environmental Law*, 1:13. Manchester University Press, Manchester.

Sanford, L. (1998) The Precautionary principle and corporate disclosure. (www.enviroweb.org/gnp/precaution)

Saunders, P.T. (2000) Use and abuse of the precautionary principle. (www.biotech-info.net/precautionary)

Scher, P., USTR Ambassador (1999) Bilateral and multilateral agricultural issues. Speech at the London School of Economics, 20 May.

Scherlen, R.G. (1998) NAFTA and beyond: the politics of trade in the post-Cold War period. In: Scott, J.M. (ed.) *After the End: Making U.S. Foreign Policy in the Post-Cold War World.* Duke University Press, Durham, North Carolina, pp. 358–385.

Schmid, G. (1989) *Report on the Proposal for a Council Directive on the Deliberate Release to the Environment of Genetically Modified Organisms.* COM (88) 160 final – SYN 131, DOC. – EN/PR/65098.TO PE 128.472/fin 28.04.1989. European Parliament Committee on the Environment, Public Health and Consumer Protection, Brussels.

Scholte, J.A., O'Brien, R. and Williams, M. (1998) *The WTO and Civil Society.* Centre for the Study of Globalisation and Regionalisation Working Paper No. 14, CSGR, Warwick, UK.

Schwarz, S. (1999) Personal interview. Foreign Agricultural Service, United States Department of Agriculture, February.

Scientific Committee for Animal Nutrition (1996) Report of the SCAN on the Safety for Animals of Certain Genetically Modified Maize Lines Notified by Ciba-Geigy in Accordance with Directive 90/220 for Feedingstuff Use. Expressed on 13 December 1996. London.

Scientific Committee for Food (1996) Opinion of the Potential for Adverse Health Effects from the Consumption of Genetically Modified Maize. Expressed on 13 December 1996. London.

Sclove, R.E. (1996) Town meetings on technology. *Technology Review.*

Scott, J. (1998) On kith and kine: trade and the environment in the EU and the WTO. *WP.*

Secretariat of the Convention on Biodiversity (2000) *Cartagena Protocol on Biosafety to the Convention on Biological Diversity: Text and Annexes.* Secretariat, Montreal. (www.biodiv.org)

Shamsie, Y. (2000) *Engaging with Civil Society: Lessons from the OAS, FTAA and Summits of the Americas.* The North–South Institute, Ottawa.

Sheppard (1997) *Spilling the Beans.* Greenpeace Discussion Paper and the Genetics Forum, London.

Shimoda, S. (1997) The biotechnology-driven transformation of agriculture. Contributed Paper at the XXII Conference of IAAE. IAAE, Sacramento, California.

Shiva, V. (1985) *Monocultures of the Mind: Perspectives on Biodiversity and Biotechnology.* Third World Network, Penang, Malaysia.

Siebert, H. (1991) Environmental policy and European integration. In: Jovanovic, M.N. (ed.) (1998) *International Economic Integration: Critical Perspectives on the World Economy – General Issues.* Vol. III. Routledge, London, pp. 748–761.

Simmons, P.J. (1998) Learning to live with NGOs. *Foreign Policy* 1998 (Autumn), 82–96.

Slovic, P. (1987) Perceptions of risk. *Science* 236(17), 280–285.

Slovic, P. (1990) The legitimacy of public perception of risk. *Journal of Pesticide Reform* 10(1), 13–14.

Sorj, B., Cantley, M. and Simpson, K. (eds) (1989) *Biotechnology in Europe and Latin America*. Kluwer Academic Publishers, Dordrecht.

Sparks Companies, Inc. (1996) Position paper to Dr M.N. Cline, Monsanto, USA, 10 September.

Spriggs, J. and Isaac, G.E. (2001) *Food Safety and International Competitiveness: the Case of Beef*. CAB International, Wallingford, UK.

Stairs, D. (2000) *Foreign Policy Consultations in a Globalizing World: The Case of Canada, the WTO and the Shenanigans in Seattle*. Policy Matters 1(8). Institute for Research on Public Policy, Ottawa.

Strange, S. (1986) *Casino Capitalism*. Basil Blackwell, Oxford.

Strange, S. (1988) *States and Markets: an Introduction to International Political Economy*. Pinter, London.

Strange, S. (1991) An eclectic approach. In: Murphy, C. and Tooze, R. (eds) *The New International Political Economy*. Lynne Reinner, Boulder, and Macmillan, London, pp. 33–49.

Swanson, T. (1997) *Global Action for Biodiversity*. Earthscan Publications, London.

Sylvander, B. and Le Floc'h-Wadel, A. (2000) Consumer Demand and Production of Organics in the EU. *AgBioForum* 3(23). (www.agbioforum.org)

Tait, J. and Levidow, L. (1992) Proactive and reactive approaches to risk regulation: the case of biotechnology. *Futures*, April, pp. 219–231.

The Express (18 September 1999) Government admits illegal GM tests. J. Ingham.

The Express (18 September 1999) Sound taste in food.

Theodorakopolou, I. and Kalaitzandonakes, N. (1999) Returns and structure of public–private knowledge networks in plant biotechnology: an EU–US comparison. Paper presented at the International Consortium on Agricultural Biotechnology Research (ICABR) 1999 Conference, The Shape of the Coming Agricultural Biotechnology Transformation: Strategic Investment and Policy Approaches from an Economic Perspective, University of Tor Vergata, Rome, Italy, 17–18 June 1999.

The Times (1999) Meacher admits GM crops illegal. V. Elliot.

Tickner, J. (1997) Precautionary principle. *The Networker* 2.

Tickner, J. (2001) An example of the precautionary principle at work: endocrine disruption. (www.islandpress.org/ecocompass/prevent)

Tickner, J. and Raffensperger, C. (2001) The precautionary principle in action: a handbook. (www.biotech-info.net/precautionary)

Tirole, J. (1988) *The Theory of Industrial Organization*. MIT Press, Cambridge, Massachusetts.

Toronto Star (1998) Public prefers genetically modified foods to be clearly labelled. 2 June.

Transatlantic Consumers Dialogue (1998) Comments on the Transatlantic Economic Partnership (TEP) Action Plan, 11 December 1998. (www.TACD.org/documents.html)

Transatlantic Consumers Dialogue (1999a) Recommendations on food safety standards, 23–24 April, Brussels. (www.TACD.org/meeting1/food.html)

Transatlantic Consumers Dialogue (1999b) EU–US summit rejects environment and consumer groups participation, invites business representatives. Press release, 18 June. (www.TACD.org/press-releases/meeting260499.html)

Transatlantic Economic Partnership (1998) Action plan (4 November 1998) *International Trade Reporter* 15(44), 1897 –1904.

United Nations (1995) World Summit for Social Development, A/CONF.166/4. United Nations, Geneva.

van den Daele, W., Puhler, A. and Sukopp, H. (1997) *Transgenic Herbicide-Resistant Crops: A Participatory Technology Assessment.* Summary Report for the Federal Ministry for Research and Technology, Wissenschaftszentrum Berlin fur Sozialforschung, Berlin.

van Scherpenberg, J. and Thiel, E. (1998) Introduction. In: van Scherpenberg, J. and Thiel, E. (eds) *Towards Rival Regionalism? US and EU Regional Regulatory Regime Building.* Nomos Verlagsgesellschaft, Baden-Baden, pp. 9–18.

Wang, Y. and Caswell, J.A. (1997) Comparative levels of food safety regulations in three U.S.–Asian trading groups. *Agribusiness* 16, 115–128.

Wartburg, W. and Liew, J. (1999) *Gene Technology and Social Acceptance.* University Press of America, New York.

WCED (World Commission on Environment and Development) (1987) *Our Common Future.* Oxford University Press, New York.

Weatherspoon, D.D., Oehmke, J.F., Wolf, C.A., Naseem, A., Maredia, M. and Hightower, A. (1999) Global Implications From a North–North–South Trade Model: a Biotech Revolution. Paper presented at the International Consortium on Agricultural Biotechnology Research (ICABR) 1999 Conference, The Shape of the Coming Agricultural Biotechnology Transformation: Strategic Investment and Policy Approaches from an Economic Perspective, University of Tor Vergata, Rome, Italy. 17–18 June 1999

Wheale, A. and McNally, R. (eds) (1990) *The BioRevolution: Cornucopia or Pandora's Box?* Pluto Press, London.

Wheale, A. and Williams, A. (1993) Between ecology and economy? The single market and the integration of environmental policy. In: Judge, D. (ed.) *A Green Dimension for the European Community: Political Issues and Processes.* Frank Cass, London, pp. 45–64.

Whitman, M. von N. (1984) Persistent unemployment: economic policy perspectives in US and Europe. In: Pierre, A. (ed.) *Unemployment and Growth in the Western Economies.* Council on Foreign Relations, New York.

WHO/FAO (World Health Organisation/Food and Agriculture Organization) (1991) *Strategies for Assessing the Safety of Foods Produced Through Biotechnology.* Geneva.

Wiegele, T.C. (1991) *Biotechnology and International Relations: the Political Dimensions.* University of Florida Press, Gainesville, Florida.

Wigan, G. (1998) Transatlantic trade issues. Speech at the London School of Economics, International Economic Diplomacy Seminar Series, November.

Winters, L.A. (1994) The EC and protection: the political economy. *European Economic Review* 38, 596–603.
Wolfe, R. (2000) *Confronting the Aftermyth of Seattle: Canada Must Set Key Trade Priorities Now.* Policy Matters 1(1), Institute for Research on Public Policy, Ottawa.
Woolcock, S. (1996) Competition among rules in the single European market. In: McCahery, J., Bratton, W.W., Picciotto, S. and Scott, C. (eds) *International Regulatory Competition and Coordination: Perspectives on Economic Regulation in Europe and the United States.* Clarendon Press, Oxford, pp. 289–321.
Woolcock, S. (1998) European and North American approaches to regulation: continued divergence? In: van Scherpenberg, J. and Thiel, E. (eds) *Towards Rival Regionalism? US and EU Regional Regulatory Regime Building.* Nomos Verlagsgesellschaft, Baden-Baden, pp. 257–276.
World Bank (1998) *World Development Report 1998/99: Knowledge for Development.* Oxford University Press, New York.
World Food Law Monthly (1999a) Thailand brings in ban over GM concerns. 19 November. AgraEurope, London.
World Food Law Monthly (1999b) 1% GM threshold backed by EU Foods Committee. 19 November. AgraEurope, London.
WTO (World Trade Organization) (1995) *Regionalism and the World Trading System.* WTO Secretariat, Geneva.
WTO (World Trade Organization) (1998a) *WTO Appellate Decision: EC – Measures Concerning Meat and Meat Products (Hormones) – Complaint by the United States and Canada.* AB-1997-4, WTO AB WT/DS26/R/USA and WT/DS48/R/Can, Geneva. (www.wto.org)
WTO (World Trade Organization) (1998b) *Appellate Decision: Australia – Measures Affecting Importation of Salmon – Complaint by Canada.* AB-1998-5, WTO AB WT/DS18, Geneva. (www.wto.org)
WTO (World Trade Organization) (1998c) *Focus, No. 31.* WTO Secretariat, Geneva. June.
WTO (World Trade Organization) (1998d) *Committee on Technical Barriers to Trade: European Council Regulation No. 1139/98 Compulsory Indication of the Labelling of Certain Foodstuffs Produced from Genetically Modified Organisms: Submission by the United States.* G/TBT/W/94 16 October 1998 98–3740, Geneva.
WTO (World Trade Organization) (1999a) *Trade and the Environment.* Special Studies 4, Geneva.
WTO (World Trade Organization) (1999b) *WTO Appellate Decision: Japan – Measures Affecting Agricultural Products – Complaint by the United States.* AB-1998-8, WTO AB WT/DS76, Geneva. (www.wto.org)
Yeutter, C.K. and Maruyapma, W.H. (1995) A NAFTA for Europe. *Wall Street Journal Europe*, 24 May. Paris.
Zampetti, A.B. (2000) Mutual recognition in the transatlantic context: some reflections on future negotiations. In: Marvoidis, P. and Cottier, T. (eds) *Regulatory Barriers and the Principle of Non-Discrimination.* University of Michigan Press, Ann Arbor, Michigan.
Zepeda, L., Douhitt, R. and You, S.-Y. (1999) Consumer acceptance of agricultural biotechnology: the role of labelling and risk perceptions on food

demand. Paper presented at the International Consortium on Agricultural Biotechnology Research (ICABR) 1999 Conference, The Shape of the Coming Agricultural Biotechnology Transformation: Strategic Investment and Policy Approaches from an Economic Perspective, University of Tor Vergata, Rome, Italy, 17–18 June 1999.

Zilberman, D. (1999) Value and pricing of agricultural biotechnology products under alternative regimes of intellectual property rights. Paper presented at the International Consortium on Agricultural Biotechnology Research (ICABR) 1999 Conference, The Shape of the Coming Agricultural Biotechnology Transformation: Strategic Investment and Policy Approaches from an Economic Perspective, University of Tor Vergata, Rome, Italy, 17–18 June 1999.

Zilberman, D., Yarkin, C. and Heiman, A. (1997) Agricultural biotechnology: economic and international implications. Contributed paper at the XXII Conference of IAAE. IAAE, Sacramento, California.

Zilberman, D., Yarkin, C. and Heiman, A. (2000) Knowledge management and the economics of agricultural biotechnology. In: Santaniello, V., Evenson, R.E., Zilberman, D. and Carlson, G.A. (eds) *Agriculture and Intellectual Property Rights: Economic, Institutional and Implementation Issues in Biotechnology.* CAB International, Wallingford, UK. Paper originally presented at the International Consortium on Agricultural Biotechnology Research (ICABR) 1999 Conference, The Shape of the Coming Agricultural Biotechnology Transformation: Strategic Investment and Policy Approaches from an Economic Perspective, University of Tor Vergata, Rome, Italy, 17–18 June 1999.

Index